How to . . .

get the most from your
COLES NOTES

Key Point

Basic concepts in point form.

Close Up

*Additional hints, notes, tips
or background information.*

Watch Out!

*Areas where problems
frequently occur.*

Quick Tip

*Concise ideas to help you
learn what you need to know.*

Remember This!

*Essential material for
mastery of the topic.*

How to get an *A* in . . .

Permutations, Combinations & Probability

Counting the outcomes

Calculating probability

Sample problems &

full solutions

ABOUT COLES NOTES

COLES NOTES have been an indispensable aid to students on five continents since 1948.

COLES NOTES now offer titles on a wide range of general interest topics as well as traditional academic subject areas and individual literary works. All COLES NOTES are written by experts in their fields and reviewed for accuracy by independent authorities and the Coles Editorial Board.

COLES NOTES provide clear, concise explanations of their subject areas. Proper use of COLES NOTES will result in a broader understanding of the topic being studied. For academic subjects, Coles Notes are an invaluable aid for study, review and exam preparation. For literary works, COLES NOTES provide interesting interpretations and evaluations which supplement the text but are not intended as a substitute for reading the text itself. Use of the NOTES will serve not only to clarify the material being studied, but should enhance the reader's enjoyment of the topic.

© Copyright 2004 and Published by

COLES PUBLISHING. A division of Prospero Books

Toronto - Canada

Printed in Canada

Cataloguing in Publication Data

Erdman, Wayne, 1954–

How to get an A in—permutations, combinations & probability

ISBN 0-7740-0571-8

1. Permutations - Problems, exercises, etc.
2. Combinations - Problems, exercises, etc.
3. Probabilities - Problems, exercises, etc.

QA164.E72 1998   511'.64   C98-930454-X

Publisher: Nigel Berrisford

Editing: Paul Kropp Communications

Book design and layout: Karen Petherick, Markham, Ontario

Manufactured by Webcom Limited

Cover finish: Webcom's Exclusive DURACOAT

Contents

PROBABILITY

Why we study permutations, combinations and probability

Mathematics, by its nature, lends itself to many applications which students will encounter when they enter work, continue study at university, and deal with other kinds of problems and interests in life. The field of permutations, combinations and probability (sometimes called Combinatorics) has wide-ranging applications, due to its basic root - counting, and its development of systematic ways of counting. These techniques develop problem-solving models for complex counting situations encountered in business, statistics, science and engineering.

An insurance company calculates life insurance premiums. An urban planner investigates the number of possible routes between two locations. An airline considers its flight network between airports. A geneticist studies the traits of a segment of the population. A biochemist researches the effectiveness of a new drug. A computer programmer designs possible routes in a new computer game. All of these require techniques from permutations, combinations and probability.

Of course, the most visible use of Combinatorics is in gambling. Successful gamblers must understand the odds of winning in any situation, even as the situation changes. While mastery of combinatorics may not lead to immediate success at a casino, it might at least ensure that you lose money more slowly than those more foolishly gambling around you.

This book uses practical examples to develop the problem-solving skills and the techniques of counting found in permutations, combinations and probability. It also develops mathematical models to be used as a reference by students when solving similar problems in their senior high school program.

Other Coles Notes covering topics in senior mathematics:

Topics in senior math are frequently interconnected at each grade level. These additional titles from Coles Notes will help you master them all:

How to Get an A in ...

- Trigonometry and Circle Geometry
- Senior Algebra
- Sequences and Series
- Statistics and Data Analysis
- Calculus

CHAPTER ONE

The fundamental counting principle

EXAMPLE 1

The location on a map is given by a letter from A to K followed by a digit from 1 to 8.

*Because there are 11 possible letters and 8 possible digits,
there are **11 x 8 = 88** possible map locations.*

This example illustrates the Fundamental Counting Principle:

Fundamental Counting Principle

If an action can be done in b ways, and a second action can be done in c ways, then the total number of ways both can be done is $b \times c$.

The following example illustrates the Fundamental Counting Principle using a tree diagram.

EXAMPLE 2

A car dealership offers a "one price deal" on a particular automobile. It can be a sedan (S) or wagon (W); it can have a power window package (P) or air conditioning (A); and a CD (C), sunroof (SR) or leather seats (L) can be added.

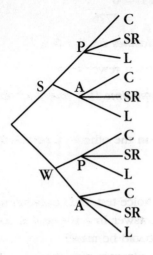

There are 3 sets of branches in this tree diagram. Each set of branches represents a set of choices for the car. If we count the number of final branches or paths, we would see that there are 12 possible combinations for this particular automobile.

By the **Fundamental Counting Principle**, 2 x 2 x 3 = 12

When developing a ***tree diagram***, remember that the branches must expand outward from the previous branch. Each set of branches represents the possible actions that can be taken. In the automobile example, there are 3 sets of actions and thus 3 sets of branches. In the map example, there would be a set of 11 branches, each connected to a set of 8 branches, for a total of 88 final branches or paths.

Frequently, an experiment can include repeated trials of the same event. The technique does not change in a major way.

If an experiment has b possible outcomes and is repeated n times (n trials), the total number of outcomes is b^n.

EXAMPLE 3

A die is rolled 5 times. How many possible outcomes can occur?

A die has 6 faces and there are 5 trials, or rolls of the dice, in this case. Therefore, we can calculate the number of outcomes as:

$$6 \times 6 \times 6 \times 6 \times 6 = 6^5$$
$$= 7776$$

∴ *There are 7776 possible outcomes on the dice.*

This same example could have been reworded as:

"Five dice are rolled at the same time. How many possible outcomes can occur?"

Because each die has no effect on the others' outcomes, the solution would be the same as above.

EXAMPLE 4

A teacher makes up a multiple-choice test. Each question has either part a, b, c or d as the correct answer. In a 10-question test, how many different answer patterns could be made?

Each question has 4 possible outcomes and there are 10 trials.

$$4^{10} = 1\ 048\ 576$$

∴ *1 048 576 different answer patterns could be made.*

EXAMPLE 5

A license plate consists of 4 letters followed by 3 digits. If there are no further restrictions, determine the total number of different license plates that can be made before a province will have to add another letter or number.

There are 7 possible positions. There are 26 choices (of letters) for each of the first 4 positions and 10 choices (of digits) for each of the next 3 positions.

$$26^4 \times 10^3 = 456\ 976\ 000$$

∴ *456 976 000 different license plates could be made.*

PRACTICE EXERCISE 1

1. A gym teacher runs a warm-up activity at the beginning of class. It goes like this:
    The first movement is either left or right.
    The second movement is forward, backward or a spin.
    The third movement is up or down.
    (a) Draw a tree diagram illustrating this activity.
    (b) How many different 3-move activities can be made?

2. A standard deck of 52 cards is cut a total of 4 times. How many outcomes of 4 cards are possible?

3. A large room has a bank of 6 light switches. Each switch could be on or off. In how many different ways could the lights be on or off?

4. When Goldilocks visited the Three Bears, the porridge was too hot, too cold and just right; the chairs were too big, too small and just right and the beds were too hard, too soft and just right. In how many ways could Goldilocks have selected her porridge, chair and bed?

5. A game is played with a 4-function calculator. In the game, a digit is pressed, followed by a function. The next person repeats the process. This is done for a total of 5 rounds, but with the last entry being an equal sign. How many arrangements of digits and functions are possible?

CHAPTER TWO

Permutations or arrangements

A **permutation** is an arrangement of elements whereby, if an element is selected, it cannot be selected again. In other words, no repetition of elements is allowed.

EXAMPLE 1

If 3 people (Corey, Sarah and Robyn) are to stand in a line for a photograph, how many arrangements could be made?

The following set of ordered triples illustrates the 6 possibilities:

(Corey, Sarah, Robyn)
(Corey, Robyn, Sarah)
(Sarah, Corey, Robyn)
(Sarah, Robyn, Corey)
(Robyn, Sarah, Corey)
(Robyn, Corey, Sarah)

There are 3 choices for the first position.

For each of those 3 choices, there are 2 choices for the second position because the first person cannot be reused.

At that point, there is only 1 choice remaining.

Multiply: 3 x 2 x 1 = 6

∴ *There are 6 possible arrangements of these people.*

The words *for each* are important in this example. They indicate that all of these people are to be used at the same time. As a result, we must multiply the numbers.

Equally important is the declining number of choices. This is because the person (or element) may not be reused.

EXAMPLE 2

Determine the number of permutations of the letters {A,B,C,D}.

Since there are 4 elements, with repetition not permitted, we multiply as follows:

$$4 \times 3 \times 2 \times 1 = 24$$

∴ *There are 24 permutations of these letters.*

Factorial notation

The number of permutations of
n elements is:

$$n! = n(n - 1)(n - 2)(n - 3) \ldots \times 3 \times 2 \times 1$$

$n!$ is read as "n factorial"

EXAMPLE 3

A child has a red, green, blue, black, orange and purple candy. In how many ways could they be lined up on a table?

$$6! = 6 \times 5 \times 4 \times 3 \times 2 \times 1$$

$$= 720$$

∴ *The candies can be lined up in 720 ways.*

EXAMPLE 4

Using the colours in Example 3, in how many ways could **only** 4 of the candies be lined up on a table?

Again, we multiply by declining values. However, there are only 4 positions, so we stop after 4 factors.

$$6 \times 5 \times 4 \times 3 = 360$$

$\therefore$ *There are 360 ways to line them up.*

To develop a general formula, let's look at this solution in factorial form, along with a parallel algebraic solution.

Allow $n = 6$ and $r = 4$. So, $n - r + 1 = 3$ and $n - r = 2$.

$$6 \times 5 \times 4 \times 3$$
$$= \frac{6 \times 5 \times 4 \times 3 \times 2 \times 1}{2 \times 1}$$

$$= \frac{6!}{2!}$$

$$n(n-1)(n-2)\ldots(n-r+1)$$
$$= \frac{n(n-1)(n-2)\ldots(n-r+1)(n-r)(n-r-1)\ldots 3 \times 2 \times 1}{(n-r)(n-n-1)\ldots 3 \times 2 \times 1}$$

$$= \frac{n!}{(n-r)!}$$

Permutations of r elements

The number of permutations of r elements taken from a sent of n elements is:

$$P(n,r) = \frac{n!}{(n-r)!}$$

The symbol $P(n,r)$ is used here. Some textbooks use $_nP_r$

EXAMPLE 5

There are 8 books in a box. Five of them are to be placed onto a shelf. In how many ways could they be arranged?

It is evident that $n = 8$ and $r = 5$.

$$P(8,5) = \frac{8!}{(8-5)!}$$

$$= \frac{8!}{3!}$$

$$= \frac{40\ 320}{6}$$

$$= 6720$$

∴ *There are 6720 arrangements of 5 books.*

EXAMPLE 6

From a standard deck of 52 cards, in how many ways could each of the following be arranged? Leave the solution in factorial form.

(a) Five face cards (J,Q,K of 4 suits).

(b) Seven hearts.

(c) Twelve red cards (hearts and diamonds).

$$\text{(a) } P(12,5) = \frac{12!}{(12 - 5)!}$$

$$= \frac{12!}{7!}$$

$$\text{(b) } P(13,7) = \frac{13!}{6!}$$

$$\text{(c) } P(26,12) = \frac{26!}{14!}$$

Note that you must not reduce the expressions, since each number represents a factorial, not a fraction.

EXAMPLE 7

Simplify each of the following.

(a) 7 x 6! $7 \times 6! = 7 \times 6 \times 5 \times 4 \times 3 \times 2 \times 1$
              $= 7!$

(b) $(n + 1)n!$ $(n + 1)n! = (n + 1)n(n - 1)(n - 2) \ldots 2 \times 1$
              $= (n + 1)!$

(c) 0!   Consider the following: $P(n,n) = \dfrac{n!}{0!}$

But $P(n,n) = n!$ (by definition of a permutation)

$$\therefore n! = \frac{n!}{0!} \qquad \therefore 0! = 1$$

8

(d) $P(n + 2, 4)$

$$P(n + 2, 4) = \frac{(n + 2)!}{(n + 2 - 4)!}$$

$$= \frac{(n + 2)!}{(n - 2)!}$$

$$= \frac{(n + 2)(n + 1)n(n - 1)(n - 2)(n - 3) \ldots 2 \times 1}{(n - 2)(n - 3) \ldots 2 \times 1}$$

$$= (n + 2)(n + 1)n(n - 1) \quad \text{after cancelling}$$

(e) $P(n + 1, n - 1)$

$$P(n + 1, n - 1) = \frac{(n + 1)!}{[(n + 1) - (n - 1)]!}$$

$$= \frac{(n + 1)!}{2!}$$

EXAMPLE 8

In how many ways could a club executive, consisting of a president, vice president and treasurer, be selected from a committee of 5 males and 5 females if:
(a) there are no further restrictions?
(b) the president and vice president may not be of the same sex?

(a) $$P(10,3) = \frac{10!}{7!}$$

$$= 720$$

∴ *There are 720 ways for the executive to be selected.*

(b) Often, it is useful to use a series of boxes to illustrate a permutation.

P	VP	T
10	5	8

There are 10 choices for the president.
Because the vice president may not be of the same sex, there are only 5 choices for that position.

9

Of the 10 original members, there are only 8 remaining.

$$10 \times 5 \times 8 = 400$$

∴ *There are 400 ways to select the executive.*

PRACTICE EXERCISE 2

1. Evaluate each of the following expressions.
    (a) P(7,3)   (b) P(12,5)

2. Simplify the following expressions.
    (a) $(n + 2)(n + 1)!$ (b) P(n,3)

    (c) $\dfrac{(n - 2)!}{(n - 5)!}$ (d) $\dfrac{P(n + 1,4)}{P(n - 1,2)}$

3. In how many ways could the 12 players on a soccer team line up for their medals if:
    (a) there are no restrictions as to order?
    (b) the captain must be first?

4. A lottery ticket contains 5 different numbers between 1 and 15, plus a bonus number. How many different lottery tickets could be printed if the order of the numbers is important and:
    (a) if the bonus number can match one of the other 5 numbers?
    (b) if the bonus number must not match one of the other 5 numbers?

5. How many 4-digit numbers could be made from the digits 1 to 7 if:
    (a) repetition of the digits is not permitted?
    (b) repetition of the digits is permitted?
    (c) repetition of the digits is not permitted and the number must be less than 4000?

6. How many arrangements are there of the letters in the word COMPUTER, if all the vowels must come first?

CHAPTER THREE

Permutations with like elements

Until now, all of the permutations have been with unlike elements. At times, some sets will have like elements, such as 3 dimes, 4 quarters and a nickel. The following examples will develop the appropriate formula needed for calculating the number of permutations.

EXAMPLE 1

How many permutations are there of 1 blue and 2 red blocks?

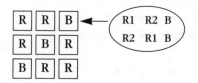

If we were to label the red blocks as R1 and R2, there would be 3! arrangements. For each arrangement of {R R B}, only 1 out of every 2! arrangements of {R,R} needs to be counted. So, we need to divide by 2!.

$$\frac{3!}{2!} = 3$$

∴ *There are 3 possible arrangements of 1 blue and 2 red blocks.*

EXAMPLE 2

How many permutations are there of 1 blue and 3 red blocks?

If we were to label the red blocks as R1, R2 and R3, there would be 3! = 6 arrangements of these 3 blocks. Only 1 out of every 3! such arrangements needs to be counted because, in reality, we cannot tell them apart. There are 4 sets of these 3! arrangements.

 ◄—

R1	R2	R3	B
R1	R3	R2	B
R2	R1	R3	B
R2	R3	R1	B
R3	R1	R2	B
R3	R2	R1	B

$$\frac{4!}{3!} = 4$$

∴ *There are 4 possible arrangements of 1 blue and 3 red blocks.*

Permutations with Like Elements

The number of permutations of *n* elements when *p, q, r, . . .* of them are alike is:

$$\frac{n!}{p!q!r! \ldots}$$

EXAMPLE 3

Determine the number of arrangements of 2 blue and 3 red blocks.

$$n = 5, \quad p = 2, \quad q = 3$$

$$\frac{5!}{2!3!} = 10$$

∴ *There are 10 arrangements of the blocks.*

12

EXAMPLE 4

In how many ways could 3 dimes, 4 quarters and a nickel be arranged in a line?

$$n = 8, \quad p = 3, \quad q = 4, \quad r = 1$$

$$\frac{8!}{3!4!1!} = 280$$

Although there is no need to include the 1! in the denominator, it is shown to make the logic clear.

∴ *There are 280 arrangements of the coins.*

EXAMPLE 5

In how many ways could 12 students be placed into 4 groups of 3 students each?

*If we were simply arranging all the students at random, there would be 12! ways of doing this. However, we must consider the fact that, within each group, order is not important. Therefore, we need to divide by the 3! arrangements of the students in **each** of the 4 groups.*

$$n = 12, \quad p = 3, \quad q = 3, \quad r = 3, \quad s = 3$$

$$\frac{12!}{3!3!3!3!} = 369\ 600$$

∴ *There are 369 600 ways of placing the students into 4 groups.*

EXAMPLE 6

(a) How many arrangements are there of the letters in the word MATHEMATICS ?
(b) How many of these arrangements begin with the letter M?
(c) How many of the arrangements in (a) would have the T's together?

(a) $n = 11, \quad p = 2, \quad q = 2, \quad r = 3$

$$\frac{11!}{2!2!2!} = 4\ 989\ 600$$

∴ *There are 4 989 600 arrangements of the letters.*

(b) There are 2 M's, one of which is placed initially, the other being used as an unlike element. As a result, there are 10 letters from which to choose.

$$n = 10, \quad p = 2, \quad q = 2$$

$$1 \times \frac{10!}{2!2!} = 907\ 200$$

There are 907 200 arrangements of the letters.

(c) We treat the two T's as one letter by "gluing" them together.

$$\therefore \quad n = 10, \quad p = 2, \quad q = 2$$

$$\frac{10!}{2!2!} = 907\ 200$$

∴ *There are 907 200 arrangements of the letters.*

EXAMPLE 7

(a) How many 6-digit numbers can be formed using only the digits 2,2,5,5,5,7?

(b) How many of these are greater than 700 000?

(a) $n = 6$ (6 *digits*), $p = 2$ (2 *twos*), $q = 3$ (3 *fives*)

$$\frac{6!}{2!3!} = 60$$

∴ 60 numbers can be formed.

(b) To be greater than 700 000, the number must begin with a 7. There are 5 remaining digits, so $n = 5$.

$$\frac{5!}{2!3!} = 10$$

∴ 10 of these numbers are greater than 700 000.

PRACTICE EXERCISE 3

1. Determine the number of arrangements of the letters in the following words:
    (a) TORONTO
    (b) ARRANGEMENT
    (c) CALCULUS

2. In question 1(c), how many of these arrangements begin with the letter C?

3. In question 1(a), how many of these arrangements have the T's together?

4. How many 7-digit numbers can be formed using only the digits 1,1,2,2,4,4,8?

5. How many 6-digit even numbers can be formed using only the digits 3,3,4,4,4,5?

6. A child is stacking 3 green, 4 blue, 1 red, 1 orange and 3 yellow blocks. In how many ways could this be done?

7. In how many ways could 12 students be divided into 3 groups of 4?

8. How many paths are possible, if you are walking 4 streets north and 5 streets west?

9. In how many ways could 3 grape, 2 orange and 5 cherry flavoured freezies be distributed among 10 children?

CHAPTER FOUR

Combinations

**A Combination is a non-ordered subset
of a set of elements.**

EXAMPLE 1

A box contains 5 marbles, all of different colours: red, green, blue, purple and yellow. Three marbles are selected at random, without regard to order. How many *different* selections of 3 marbles can be made?

(R)(G)(B) ◄—

(R)(G)(P)

(R)(G)(Y)

(R)(B)(P)

(R)(B)(Y)

(R)(P)(Y)

(R)(B)(P)

(R)(B)(Y)

(R)(P)(Y)

(R)(P)(Y)

R	G	B
R	B	G
G	R	B
G	B	R
B	R	G
B	G	R

If this were a permutation,

$$P(5,3) = \frac{5!}{2!}$$

$$= 60$$

*there would be 60
permutations of 3 colours.*

*However, there are 3! = 6
arrangements of each subset,
only one of which is counted.
This is repeated for each
selection of 3 colours. We must
divide by 3!.*

16

$$\frac{60}{3!} = 10$$

∴ *There are 10 ways of selecting 3 marbles.*

To derive the general case, we do the following:

$$\frac{P(n,r)}{r!} = \frac{\dfrac{n!}{(n-r)!}}{r!}$$

$$= \frac{n!}{(n-r)!r!}$$

Notice here, we are dividing by $r!$ to eliminate the unwanted duplication of the subsets, since order is not important.

The Number of Combinations (or Subsets)

The number of combinations of r elements taken from a set of n elements, without regard to order is:

$$C(n,r) = \frac{n!}{(n-r)!r!}$$

Some books write this as $_nC_r$ or $\begin{pmatrix} n \\ r \end{pmatrix}$

Read this as: n choose r.

EXAMPLE 2

Five children are to be chosen from a group of 9. In how many ways could this be done?

These children are selected to a non-ordered group, so we use the combinations formula.

$$n = 9, \quad r = 5$$

$$C(9,5) = \frac{9!}{4!5!}$$

$$= 126$$

This can be done in 126 ways.

EXAMPLE 3

(a) In how many ways could a hand of 5 cards be dealt from a deck of 52 cards?

(b) In how many ways could the 5 cards be face cards (J,Q,K)?

Because cards dealt to a hand are non-ordered, we use the combination formula:

(a) $n = 52, \quad r = 5$

$$C(52,5) = \frac{52!}{47!5!}$$

$$= 2\ 598\ 950$$

There are 2 598 950 ways to deal 5 cards from a deck.

(b) $n = 12, \quad r = 5$

$$C(12,5) = \frac{12!}{7!5!}$$

$$= 792$$

The 5 cards can be face cards in 792 ways.

EXAMPLE 4

In how many ways can a committee of 4 people be selected from a club with 30 members?

$$n = 30, \quad r = 4$$

$$C(30,4) = \frac{30!}{26!4!}$$

$$= 27\ 405$$

A committee of 4 can be selected in 27 405 ways.

EXAMPLE 5

In how many ways could 3 girls and 2 boys be selected from a group of 9 girls and 6 boys?

*Here, two groups are being selected at the **same time**, so we will **multiply** the results of the boys and girls together.*

Girls: n = 9,   r = 3   Boys: n = 6,   r = 2

$$C(9,3) \times C(6,2) = \frac{9!}{6!3!} \times \frac{6!}{4!2!}$$

$$= 84 \times 15$$

$$= 1260$$

∴ *They could be selected in 1260 ways.*

EXAMPLE 6

Evaluate each of the following. State a conclusion as they refer to combinations.

   (a) C(8,3)   (b) C(8,5)

$$C(8,3) = \frac{8!}{5!3!} \qquad\qquad C(8,5) = \frac{8!}{3!5!}$$

$$= 56 \qquad\qquad\qquad\qquad = 56$$

   (c) C(11,7)   (d) C(11,4)

$$C(11,7) = \frac{11!}{4!7!} \qquad\qquad C(11,4) = \frac{11!}{7!4!}$$

$$= 330 \qquad\qquad\qquad\qquad = 330$$

*Conclusion: The number of combinations of **r** elements is the same as the number of combinations of **n − r** elements.*

***EXAMPLE* 7**

Given $C(n, n - 2) = 28$, $n \in$ N, solve for n.

$$C(n, n - 2) = 28$$

$$\frac{n!}{2!(n - 2)!} = 28$$

$$\frac{n(n - 1)(n - 2)(n - 3) \ldots 1}{2!(n - 2)(n - 3) \ldots 1} = 28$$

$$\frac{n(n - 1)}{2} = 28$$

$$n(n - 1) = 56$$

$$n^2 - n - 56 = 0$$

$$(n - 8)(n + 7) = 0$$

$$n = 8 \text{ or } n = -7$$

But, $n \geq 0$
$\therefore n = 8$

PRACTICE EXERCISE 4

1. Fifteen students signed up to go on a ski trip. There is room for only 10 of them. In how many ways can the 10 students be selected?

2. How many hands of 13 cards contain:
    (a) only red cards?
    (b) 2 spades, 3 hearts, 5 diamonds and 3 clubs?
    (c) no face cards?
    Leave your answer in unsimplified form.

3. In how many ways could 15 students be divided into 3 groups of 5 students?

4. There are 17 people at a party. Each person shakes hands with all of the other people. How many handshakes occur?

5. A photographer has 10 different-colored filters.
    (a) How many combinations of 2 filters can be made?
    (b) How many combinations of 3 filters can be made?
    (c) In how many ways could 2 filters, and then 3 more, be selected from the 10 filters?

6. Evaluate each of the following:
    (a) C(12,8)   (b) C(7,0)
    (c) C(450,0)   (d) C(145,145)
    (e) C(10,3) x C(7,2)   (f) C(12,4) x C(8,4) x C(4,4)

7. Given $C(n,2) = 15$,   solve for $n \in$ N.

CHAPTER FIVE

Total number of subsets of a set

EXAMPLE 1

Given the set {a,b,c,d,e}, determine the number of subsets.

The number of subsets of 0 elements is C(5,0) = 1
The number of subsets of 1 element   is C(5,1) = 5
The number of subsets of 2 elements is C(5,2) = 10
The number of subsets of 3 elements is C(5,3) = 10
The number of subsets of 4 elements is C(5,4) = 5
The number of subsets of 5 elements is C(5,5) = 1

By adding these up, we see the total number of subsets of a set of 5 elements is 32.

Using the above example as a guide, we can develop a formula for the total number of subsets of any set. We will use a table of values for this development.

No. of elements	Sum of Number of Subsets	Total
0	C(0,0)	1
1	C(1,0) + C(1,1)	2
2	C(2,0) + C(2,1) + C(2,2)	4
3	C(3,0) + C(3,1) + C(3,2) + C(3,3)	8
4	C(4,0) + C(4,1) + C(4,2) + C(4,3) + C(4,4)	16
5	C(5,0) + C(5,1) + C(5,2) + (C5,3) + C(5,4) + C(5,5)	32

Following the pattern, we can see that the **total** column is developed as $2^0, 2^1, 2^3, 2^4, 2^5$.

This can be proven by considering a set of n elements. Each element can be either **included** or **excluded** in a given subset.

So, to form a subset, each element will have 2 choices — inclusion or exclusion.

Thus there are 2 x 2 x 2 x ... x 2 ways of forming subsets. Because there are *n* elements, there are *n* factors of 2. As a result this can be written as 2^n.

Total Number of Subsets of a Set

The total number of subsets of a set of *n* elements is 2^n.

EXAMPLE 2

There are 6 flags in a set of signal flags. One or more of them can be used, without regard to order, in forming signals. How many different signals can be made?

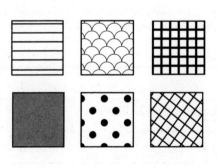

One or more flags make up a complete signal. Because at least 1 flag must always be used, we must subtract C(6,0), representing the empty set, Ø.

$$2^6 - 1 = 63$$

∴ *There are 63 different signals that can be made.*

23

EXAMPLE 3

A child wishes to buy some candy, but his mother said, "No more than one." A candy store has 10 different 5 cent candies. In how many ways could this child buy *no more than one* of each candy? (The amount of money available is not relevant.)

The child may buy either 0 or 1 of each candy thereby providing 2 choices for each.

Because the child will buy at least one candy, we must subtract 1, representing Ø.

$$2^{10} - 1 = 1023$$

∴ *The child could buy candies in 1023 ways.*

EXAMPLE 4

Determine the number of subsets of the set {a,a,b,b,c,c,d,d}.

Because there is duplication of elements, we cannot simply say that the answer is 2^8.
However, we can form our subsets in a manner similar to our original proof of the formula.
Each letter can be taken in 3 ways: 0, 1 or 2 letters included.

$$3 \times 3 \times 3 \times 3 = 3^4$$
$$= 81$$

∴ *There are 81 subsets.*

From these examples, we can generalize that, if there are **p** elements alike, then there are **p + 1** ways to include them in a subset.

Number of Subsets When Some are Alike

The number of subsets of a set of *n* elements when *p, q, r, ...* of them are alike is:

$$(p + 1)(q + 1)(r + 1) \ldots \times 2^{n-p-q-r\ldots}$$

EXAMPLE 5

A gardener has 4 red, 3 yellow, 2 white and 1 each of pink, blue and purple tulip bulbs from which to choose. In how many ways could this gardener select some bulbs?

$$n = 12, \quad p = 4, \quad q = 3, \quad r = 2$$
and we must exclude the Ø.

$$(4 + 1)(3 + 1)(2 + 1) \times 2^{12-4-3-2} - 1 = 479$$

There are 479 ways for this gardener to select some bulbs.

EXAMPLE 6

(a) Determine all the prime factors of 4200.
(b) What is the total number of factors of 4200?

  (a) $4200 = 2 \times 2 \times 2 \times 3 \times 5 \times 5 \times 7$
  (b) All factors are made up of subsets of the prime factors $\{2,2,2,3,5,5,7\}$

$$n = 7, \quad p = 3, \quad q = 2$$

$$(3 + 1)(2 + 1) \times 2^{7-3-2} = 48$$

∴ *There are 48 factors of 4200.*

Keep in mind that this includes 1 and 4200. If we were to exclude them, we would need to subtract 2 from our total.

EXAMPLE 7

A set has 32 766 subsets, excluding Ø and itself. How many elements are in the set?

Because Ø and the set itself are excluded,
we must subtract 2 from 2^n.

$$2^n - 2 = 32\ 766$$
$$2^n = 32\ 768$$
$$n = 15$$

∴ *The set has 15 elements.*

PRACTICE EXERCISE 5

1. Determine the total number of subsets of each of the following sets.
   (a) {h,i,j,k,l,m,n}
   (b) {red, green, blue, black, orange, yellow, purple, brown, pink, white}
   (c) {1,1,1,2,3,4,5,5,6}

2. The Ice Cream Boutique offers 7 different toppings for their ice cream sundaes. You may select one or more different toppings for their vanilla ice cream. How many different vanilla ice cream sundaes could be made?

3. Determine the total number of divisors of the number 51 975, excluding 1 and itself.

4. Determine the value of n:
   $C(n,n) + C(n,n - 1) + ... + C(n,0) = 8192$

5. A person has 4 pennies, 2 dimes, 3 quarters and 5 loonies. How many different sums of money could this person make?

6. A binary code is used to represent numbers and characters in computers. The code contains only 0's and 1's as digits (for example, 10110001). These are known as *binary digits*, or *bits*. How many numbers and/or characters could be formed:
   (a) using up to 8 digits?
   (b) using up to 16 digits?
   (c) using up to 32 digits?

CHAPTER SIX

Problem solving with permutations and combinations

The hardest part of problem solving is to identify whether the solution requires a permutation or combination type of solution. In addition, sifting through all the information and details to determine the values of n and r can be equally difficult. This chapter will offer some tips on overcoming these difficulties.

The first step is often to determine whether **order** is important. Example 1 illustrates this point.

EXAMPLE 1

A club contains 15 members. In how many ways could
(a) an executive of 3 members be selected?
(b) a president, vice president and treasurer be selected?

In (a) the executive members are not given titles, so order is not important, whereas in (b) the order is important. Therefore, (a) is a combination and (b) is a permutation.

(a) $C(15,3) = 455$

∴ *There are 455 ways to select an executive.*

(b) $P(15,3) = 2730$

∴ *There are 2730 ways to select a president, vice president and treasurer.*

EXAMPLE 2

Twelve points are arranged in a circle. In how many ways could they be joined together with straight lines?

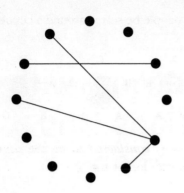

*Any 2 points form a straight line. As well, a given line goes through 2 points, **not from** one to the other. Therefore, order is not important.*

$$C(12,2) = 66$$

∴ There are 66 ways to join 2 points.

EXAMPLE 3

Identify whether or not order is important in each of the following scenerios:

(a) Selecting 6 numbers from a list of 49 numbers for a lottery ticket.
(b) Making up a seating plan for the head table at a banquet.
(c) Submitting a batting order for a baseball team.
(d) Assigning 3 students to a car and 5 to a van for a field trip.
(e) Determining the number of hearts, diamonds and spades in a hand.
(f) Determining the total number of license plates that can be made from a group of letters and numbers.

(a) *Order is not important on this **type of lottery ticket.***
(b) *A **seating plan** implies order is important.*
(c) *For a baseball team, a **batting order** implies order is important.*
(d) *The students are **not placed in a particular order** within the vehicles. Therefore, order is not important.*
(e) *The **cards in a hand** are not placed in any specific order, Therefore, order is not important.*
(f) ***Changing the order** of the numbers or letters **changes the license plate**. Therefore, order is important.*

EXAMPLE 4

In how many ways could 5 people be seated around a circular table?

Each of the above seating plans is considered to be the same because of the circular table. So we do not use 5!

The seating plan on the left is different from all the ones above.
Therefore order is important.
To overcome the "circular" problem, it is best to allow one person to be seated and arrange the other 4 around him or her.

$$1 \times 4! = 24$$

∴ *There are 24 different seating plans.*

EXAMPLE 5

In how many ways could 16 people be assigned to a 5-person car, a 4-person car and a 7-person van?

Within the vehicles, order is not important. However, as the people are selected for each vehicle, there is a declining total from which to choose: 16 - 5 = 11 and then 11 - 4 = 7.
*We **multiply** because they are being **assigned together**.*

$$C(16,5) \times C(11,4) \times C(7,7) = 1\ 441\ 440$$

∴ *There are 1 441 440 ways of seating the people.*

EXAMPLE 6

(a) In how many ways could 6 girls and 5 boys line up, if they must alternate positions?

◯ = Girl ⬛ = Boy

*Order is important because they are "**lined up**".*
There are 6! arrangements of the girls and 5! of the boys.
***Multiply** because they are lined up at the*
***same time**.*

$$6! \times 5! = 86\ 400$$

They could be lined up in 86 400 ways.

(b) In how many ways could the children line up if there were
5 girls and 5 boys?

Because there is an equal number of boys and girls,
*   we could:*
*   - start with a girl and finish with a boy; or*
*   - start with a boy and finish with a girl.*
Therefore, we must multiply by 2.

$$5! \times 5! \times 2 = 28\ 800$$

They could be lined up in 28 800 ways.

EXAMPLE 7

How many 7-digit even numbers less than 5 000 000 can be
formed using all the digits 1,2,2,3,5,5,6?

Because we are forming "numbers," order is important.
*What makes this example difficult is the **overlapping***
***restrictions**. As a result, problems like these are best done*
*using **cases**.*

Case 1: *The last digit is a 2; the first digit is a 1,2 or 3.*
One choice {2} at the end for even.
Three choices {1,2,3} remain to be less than
5 000 000.

$$\underline{3}\ \underline{\ }\ \underline{\ }\ \underline{\ }\ \underline{\ }\ \underline{1}$$

⟺

Middle has no restriction except for the double 5.

$$\frac{3 \times 1 \times 5!}{2!} = 180$$

Case 2: *The last digit is 6; the first digit is a 1 or 3.*
One choice {6} at the end for even.
Two choices {1,3} remain to be less than
5 000 000.
Middle has no restriction except for the double
2 and double 5.

$$\frac{2 \times 1 \times 5!}{2!2!} = 60$$

Case 3: *The last digit is a 6; the first digit is a 2.*
One choice {6} at the end for even.
One choice {2} at the beginning to be less
than 5 000 000.
Middle has no restriction except for the
double 5.

$$\frac{1 \times 1 \times 5!}{2!} = 60$$

Add the 3 outcomes together because they are
different alternatives:

$$180 + 60 + 60 = 300$$

∴ *There are 300 even numbers less than 5 000 000.*

EXAMPLE 8

A particular volleyball team has 2 setters, 4 middle hitters, 4 power hitters and 2 offside hitters. The team has been asked to send representatives to a clinic, based on their roles. In how many ways could the team be represented at the clinic?

The players are not being placed in a specific order.
*So, order is not important. However, between 1 **and***
***12 players** could represent the team. Therefore, we will*

*use the **total number of subsets** method, subtracting
1 because the team must be represented.*

$$(2+1)(4+1)(4+1)(2+1) \times 2^0 - 1 = 224$$

∴ There are 224 ways the team could be represented.

EXAMPLE 9

(a) In a poker hand of 5 cards, how many ways are there to get a
full house (3 of one kind, 2 of another)?

*Order is not important in hands of
cards.
Select a denomination {ace to king =
13} for the 3-of-a-kind.
There are 3 cards to be selected from
4 of that denomination.*

$$\Rightarrow C(13,1) \times C(4,3)$$

*Select another denomination from the remaining 12 for
the 2-of-a-kind.
There are 2 cards to be selected from 4 of that
denomination.*

$$C(12,1) \times C(4,2)$$

$$C(13,1) \times C(4,3) \times C(12,1) \times C(4,2) = 3744$$

∴ There are 3744 ways to get a full house.

(b) In a poker hand of 5 cards, how many ways are there to get a
royal flush (10, Jack, Queen, King, Ace of the same suit)?

*Select the suit {one of hearts, diamonds,
clubs, spades}
Then select all of the 5 named cards in
that suit.*

$$C(4,1) \times C(5,5) = 4$$

There are 4 ways to get a royal flush.

EXAMPLE 10

How many subsets of {1,1,2,2,2,3,4,4,5} will contain a 2?

The statement "will contain a 2" tells us that 1, 2 or 3 2's are in each subset. We could solve this using cases. The easier method is to take the *total without restriction* and *subtract* the number with no 2.

$$(2 + 1)(3 + 1)(2 + 1) \times 2^2 - (2 + 1)(2 + 1) \times 2^2 = 144 - 36$$
$$= 108$$

∴ *108 subsets will contain a 2.*

EXAMPLE 11

In how many ways could the 10 questions on a test be arranged, if the hardest and easiest:

(a) are side-by-side?

(b) are not side-by-side?

(a) *As is stated in the question, this is an arrangement problem.*
To place them side-by-side, we "glue" them together, treating them as one question. There will then be the equivalent of only 9 questions. We must also multiply by 2! because we must include the 2 ways of "gluing" them together {AB, BA}.

$$9! \times 2! = 725\ 760$$

∴ *There are 725 760 ways to arrange the questions.*

(b) *As in Example 10, the easiest way is to **subtract** part (a) from the **total** number of arrangements **without restriction**.*

$$10! - 9! \times 2! = 2\ 903\ 040$$

∴ *There are 2 903 040 ways to arrange the questions.*

Problem solving strategies

1. Determine if order is important. Look for hints in the problem, or use reasoning or common sense. If yes, then use a permutation model. If no, use a combination model. (Examples 1, 2, 3)

2. Look for restrictions. Take care of them first. (Examples 4, 7, 9)

3. If there is repetition that needs to be eliminated, it is usually done by division by p! if a permutation or by multiplying by $p + 1$ if finding total number of subsets. (Examples 7, 8, 10)

4. If the events occur at the same time, remember that the outcomes are multiplied. If they occur independently, at different times, add the outcomes. (Examples 5, 6, 7)

5. Diagrams are extremely useful. (Examples 2, 4, 6, 9)

6. Place holders for the elements are useful. (Example 7)

7. There may be declining totals. (Examples 5, 9)

8. You may need to find all the possible combinations. (Examples 8, 10)

9. Build a smaller model if the problem uses large numbers. (In Example 4, draw diagrams for 3 people instead of 5.)

10. If elements need to be together or apart, try "gluing" them together. Then decide if they need to be arranged amongst themselves. (Example 11)

11. When elements must be included or excluded, it may be useful to subtract the converse from the total without restriction. (Examples 10, 11)

12. Finally, try to build a solution based on a model you already know. Most problems will be similar in some way, but rarely will they be identical.

PRACTICE EXERCISE 6

1. How many "words" can be formed using all the letters in the word VANCOUVER, if the vowels must be in ascending (AEOU) order?

2. In how many ways could 9 people be seated around a circular table?

3. How many different 8-bead necklaces can be made from 8 different beads placed on a leather string that has a clasp?

4. How many diagonals does a hexagon have?

5. In how many ways can a committee, consisting of a president, secretary and 3 other members, be selected from a 16-member list?

6. If the union president and chief company representative are not to be seated together, in how many ways could the 10 members of a negotiating committee be seated around a circular table?

7. In how many ways could 20 copies of the same book be placed onto 4 shelves, if each shelf must contain at least 1 book?

8. How many 5-card poker hands contain:
   (a) a straight flush (5 consecutive cards from only 1 suit)?
   (b) a straight (5 consecutive cards from at least 2 suits)?

9. In how many ways could a committee of 5 be formed from 8 men and 6 women, if
   (a) there is no restriction?
   (b) the committee must contain at least one of each sex?

10. How many different masses can be formed from masses of 16 kg, 4 kg, 2 kg, 1 kg and 2 of 8 kg?

CHAPTER SEVEN

Extensions of combinations

Pascal's Triangle

Pascal's Triangle was attributed to Blaise Pascal (1623 - 1662), a French mathematician. It has since been discovered that the Chinese (1200's) and Egyptians (1400's) (1400's) knew of the properties of the triangle long before Pascal, but the name remains.

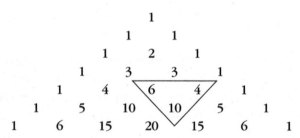

The entries are generated by adding two adjacent entries and placing the results below and between them, as indicated by the triangle drawn around 6-4-10.

Now, how does this relate to the topic of combinations? If we build a similarly shaped triangle of $C(n,r)$ as follows, the results will be obvious.

$$C(0,0)$$
$$C(1,0) \quad C(1,1)$$
$$C(2,0) \quad C(2,1) \quad C(2,2)$$
$$C(3,0) \quad C(3,1) \quad C(3,2) \quad C(3,3)$$
$$C(4,0) \quad C(4,1) \quad C(4,2) \quad C(4,3) \quad C(4,4)$$
$$C(5,0) \quad C(5,1) \quad C(5,2) \quad C(5,3) \quad C(5,4) \quad C(5,5)$$
$$C(6,0) \quad C(6,1) \quad C(6,2) \quad C(6,3) \quad C(6,4) \quad C(6,5) \quad C(6,6)$$

When we evaluate each of the C(*n,r*) entries, the results will match those in the original triangle:

C(0,0) = 1, C(3,2) = 3, C(5,3) = 10, C(6,2) = 15, etc.

Before the advent of calculators and spreadsheets, Pascal's Triangle was an invaluable tool for calculating values of C(*n,r*). There are many other uses and properties we will investigate in this chapter.

For studying purposes, it is a good idea to build your own copy of Pascal's Triangle (both versions) to about 12 rows.

EXAMPLE 1

Determine the sum of the entries in each row of Pascal's Triangle. Make a conclusion based on your results.

Whenever you are investigating patterns, including those in Pascal's Triangle, it is usually best to build a table of values. (When it involves complex or long mathematical formulas, use a spreadsheet.)

ROW	n	SUM	OBSERVATION
0	0	1	2^0
1	1	2	2^1
2	2	4	2^2
3	3	8	2^3
4	4	16	2^4
5	5	32	2^5

With careful observation of the pattern, we can see that the sum of the entries in a given row is 2^n.

This follows our previous observation in Chapter 5 that the total number of subsets of a set is 2^n.

EXAMPLE 2

In the grid shown, determine the number of paths from point A to point B, travelling on the lines and only downward or to the right.

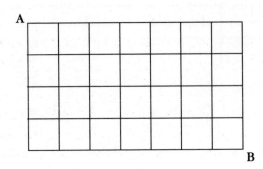

By investigating results involving smaller, but increasingly larger, grids, we can observe a pattern:

1 by 1 - 2 paths

1 by 2 - 3 paths

1 by 3 - 4 p

2 by 2 - 6 paths

Copy these numbers onto the grid as shown below:

A

2	3	4				
3	6					
4						

B

Observe the pattern in the grid and see that the entries are those of Pascal's Triangle. The entry at B could be found by adding pairs of previous terms or, since this is a 7 by 4 grid, the entry would be:

$$C(11,7) = C(11,4)$$

$$= 330$$

There are 330 paths from A to B.

THE BINOMIAL THEOREM

The Binomial Theorem is one that enables us to expand and simplify expressions such as $(2x - 3y^2)^6$ with relative ease.

Start by expanding and simplifying each of the following and investigating the pattern.

$(x + y)^0 = \mathbf{1}$

$(x + y)^1 = \mathbf{1}x + \mathbf{1}y$

$(x + y)^2 = (x + y)(x + y) = \mathbf{1}x^2 + \mathbf{2}xy + \mathbf{1}y^2$

$(x + y)^3 = (x + y)(\mathbf{1}x^2 + \mathbf{2}xy + \mathbf{1}y^2) = \mathbf{1}x^3 + \mathbf{3}x^2y + \mathbf{3}xy^2 + \mathbf{1}y^3$

$(x + y)^4 = (x + y)(\mathbf{1}x^3 + \mathbf{3}x^2y + \mathbf{3}xy^2 + \mathbf{1}y^3)$

$\qquad = \mathbf{1}x^4 + \mathbf{4}x^3y + \mathbf{6}x^2y^2 + \mathbf{4}xy^3 + \mathbf{1}y^4$

Careful observation of the coefficients (in **bold**) of the terms will show that they follow the entries in each row of Pascal's Triangle.

EXAMPLE 3

Use Pascal's Triangle to expand $(x + y)^6$.

$$(x + y)^6 = 1x^6 + 6x^5y + 15x^4y^2 + 20x^3y^3 + 15x^2y^4 + 6xy^5 + 1y^6$$

Quick observation allows us to extend this to using C(n,r) notation in expanding binomials. This is known as the Binomial Theorem.

 Binomial Theorem

$$(x + y)^n = C(n,0)x^n + C(n,1)x^{n-1}y + C(n,2)x^{n-2}y^2 + ... + C(n,n)y^n$$

EXAMPLE 4

Use the Binomial Theorem to expand and simplify $(x + y)^5$.

$(x + y)^5$

$= C(5,0)x^5 + C(5,1)x^4y + C(5,2)x^3y^2 + C(5,3)x^2y^3 + C(5,4)xy^4 + C(5,5)y^5$

$= x^5 + 5x^4y + 10x^3y^2 + 10x^2y^3 + 5xy^4 + y^5$

EXAMPLE 5

Use the Binomial Theorem to expand and simplify $(2x - 3y^2)^6$.

Rather than writing in C(n,r) form, simply use Pascal's Triangle for the coefficients. Be careful, also, to expand the brackets properly.

$(2x - 3y^2)^6$

$= (2x)^6 + 6(2x)^5(-3y^2) + 15(2x)^4(-3y^2)^2 + 20(2x)^3(-3y^2)^3 + 15(2x)^2(-3y^2)^4 + 6(2x)(-3y^2)5 + (-3y^2)^6$

$= 64x^6 - 576x^5y^2 + 2160x^4y^4 - 4320x^3y^6 + 4860x^2y^8 - 2916xy^{10} + 729y^{12}$

EXAMPLE 6

On the game board shown, a checker, O, may move forward diagonally left or right. If an X is encountered, the checker may not jump over it. Determine the number of paths to the top of the board.

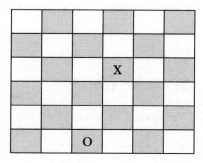

As we move up the board, there is 1 path to each of the first squares diagonally above the O. Following that, we use Pascal's method and add adjacent numbers to get further entries. However, an X blocks our path, not permitting us to add.

	6		4		1
3		3		4	
	3		X		1
1		2		1	
	1		1		
		O			

$$6 + 4 + 1 = 11$$

There are 11 paths to the top of the game board.

PRACTICE EXERCISE 7

1. On the game board shown, a checker, O, may move forward diagonally left or right. If an X is encountered, the checker may jump over it. Determine the number of paths to the top of the board.

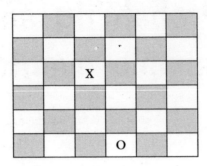

2. What are known as "triangular numbers" can be derived by counting the number of circles used to form and fill in a triangle. Continue the sequence to develop a pattern that can be identified in Pascal's Triangle. Generalize with a formula.

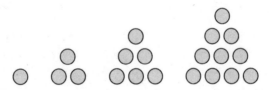

3. Use the Binomial Theorem to expand and simplify each of the following.

(a) $(4x - y)^6$ (b) $(x^2 + 3y)^4$ (c) $\left(2x + \dfrac{1}{x^2}\right)^5$

CHAPTER EIGHT

Probability and simple events

Probability, as an area of study, is the mathematics of chance.

Probability, as a numerical value, is calculated as:

$$P = \frac{n(E)}{n(S)} = \frac{\text{the number of successful outcomes in an event}}{\text{the total number of outcomes in an experiment or sample space}}$$

Essentially, there are two forms of probability: experimental and theoretical. Experimental probability relies on a large number of observations in an experiment in order to reach an accurate calculation. An example of this would be accident results for calculating automobile insurance rates. Theoretical probability relies on a mathematical model based on all the possible outcomes, or sample space.

Based on the fact that the number of successful outcomes is always less than or equal to the total number of outcomes, it is correct to say that the probability of an event occurring is between 0 and 1. This can be interpreted in the following diagram:

```
        Unlikely              Likely
   ┌───────────────┬───────────────┐
   0              0.5              1
   Impossible                   Certain
```

EXAMPLE 1

(a) Tossing heads with a quarter would have a probability of 0.5.

(b) Tossing 3 heads on 2 coins would have a probability of 0.

(c) Tossing a head or tail on a coin would have a probability of 1.

(d) Tossing at least one head on 3 coins would be likely.

(e) Tossing 3 heads on 3 coins would be unlikely.

EXAMPLE 2 (Experimental probability)

During a recent survey of ethnic backgrounds of 1000 people in a large city, 513 were British, 148 were French, 72 were African, 56 were Chinese and the remainder were from other groups.

Calculate the probability that a person, selected from the population at random has:

(a) a British background.

(b) an African background.

(c) an "other" background.

(a) $P = \dfrac{513}{1000} = 0.513$ *(Likely)*

(b) $P = \dfrac{72}{1000} = 0.072$ *(Unlikely)*

Subtract the probability of the given ethnic groups from 1 (total).

(c) $P = 1 - \dfrac{513 + 148 + 72 + 56}{1000} = 0.211$ *(Unlikely)*

The above example is experimental in nature because the numbers were developed from a large database of 1000 responses, not the entire population.

EXAMPLE 3 (Theoretical probability)

A card is drawn at random from a standard deck of 52 cards. Determine the probability that:

(a) it is a king;

(b) it is red;

(c) it is a face card (J, Q, K).

(a) $P = \dfrac{4}{52}$ *(Unlikely)*

(b) $P = \dfrac{26}{52} = \dfrac{1}{2}$ *(Even)*

(c) $P = \dfrac{12}{52} = \dfrac{3}{13}$ *(Unlikely)*

The above example is theoretical in nature because it uses the whole sample space as the basis for its calculations. Examples showing equally likely outcomes:

Uniform Probability Distribution

In a Uniform Probability Distribution, each outcome is equally likely.

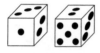

Dice

Raffle or lottery tickets

EXAMPLE 4

Show the complete distribution of probabilities for the spinner in the above diagram.

OUTCOME	PROBABILITY
1	0.125
2	0.125
3	0.125
4	0.125
5	0.125
6	0.125
7	0.125
8	0.125

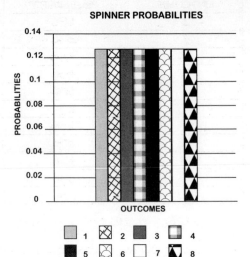

SPINNER PROBABILITIES

Legend: 1, 2, 3, 4, 5, 6, 7, 8

EXAMPLE 5

Two dice are rolled and the sum of the upper faces is found. Determine the probability of each sum.

SUMS OF 2 DICE

		1	2	3	4	5	6	Die 2
Die 1	1	2	3	4	5	6	7	
	2	3	4	5	6	7	8	
	3	4	5	6	7	8	9	
	4	5	6	7	8	9	10	
	5	6	7	8	9	10	11	
	6	7	8	9	10	11	12	

Each entry in the table is equally likely, so the entries could be represented by a uniform distribution, each with a probability of 1/36. However, each sum is not equally likely. The probability distribution is shown on the next page:

OUTCOME	PROBABILITY
2	1/36
3	2/36
4	3/36
5	4/36
6	5/36
7	6/36
8	5/36
9	4/36
10	3/36
11	2/36
12	1/36

If we add the probabilities in this or any distribution, we would find the sum to be 1.

Mathematical Expectation

Mathematical Expectation is the outcome that can be expected to be obtained, when probability is factored in. In statistics, it is called a mean or average. To calculate the expectation, multiply the individual outcomes with their probabilities, then add.

$$E(x) = \sum x P(x)$$

where x is the individual outcome and $P(x)$ is its probability.

EXAMPLE 6

Determine the mathematical expectation of the spinner in Example 4.

OUTCOME x	PROBABILITY $p(x)$	xP(x)
1	0.125	0.125
2	0.125	0.25
3	0.125	0.375
4	0.125	0.5
5	0.125	0.625
6	0.125	0.75
7	0.125	0.875
8	0.125	1
Sum		4.5

$$E(x) = \Sigma xP(x) = 4.5$$

∴ *The expectation of the spinner is 4.5.*

This means that, over an extended period of time, the average spinner result would be 4.5. (Note that it is not simply 1/2 of 8 because the midpoint of the list is halfway between 4 and 5, or 4.5.)

Odds

Given that *p* is the probability of the event occurring and q is the probability that it will not occur, (*q = 1 − p*):

The odds *against* an event occurring are the ratio *q:p*.
The odds *in favour of* an event occurring are the ratio *p:q*.

EXAMPLE 7

The Winnipeg hockey club, the Wolves, is holding its annual lottery: 20 000 tickets have been printed and sold. Calculate the odds **against** winning first prize.

Each ticket has an equally likely chance of winning.

$$p = \frac{1}{20\ 000} \qquad q = 1 - \frac{1}{20\ 000}$$

$$= \frac{19\ 999}{20\ 000}$$

$$Odds\ against = \frac{19\ 999}{20\ 000} : \frac{1}{20\ 000}$$

$$= 19\ 999 : 1$$

Be careful here: a common mistake is to state the odds against as 20 000:1.

EXAMPLE 8

An integer between 1 and 25 inclusive is selected at random. Determine the odds in favour of it

(a) being even.

(b) being a perfect square.

(c) being greater than 10.

(a) $p = \frac{12}{25} \Rightarrow q = \frac{13}{25}$

Odds in favour = 12:13

(b) $p = \frac{5}{25} \Rightarrow q = \frac{20}{25}$

Odds in favour = 5:20 = 1:4

(c) $p = \frac{15}{25} \Rightarrow q = \frac{10}{25}$

Odds in favour = 15:10 = 3:2

PRACTICE EXERCISE 8

1. A coin is tossed twice. Determine the probability that
   (a) two tails show.
   (b) a head and then a tail show.
   (c) a head and a tail show.

2. For a school lottery, 1250 tickets were sold. There is a grand prize as well as 2 second prizes and 5 third prizes. Determine the odds in favour of:
   (a) winning a prize.
   (b) winning the grand prize.
   (c) winning a second or third prize.

3. A bag contains 2 red, 3 green, 1 blue and 5 white balls. A child reaches into the bag and randomly selects 1 ball. Construct a table to show the complete distribution of probabilities for all the possible outcomes.

4. If, in question 3, the first ball is replaced and a ball is selected again:
   (a) Make a tree diagram to illustrate all the possible outcomes of the child's two draws.
   (b) Calculate the probability for each outcome.

5. One year, oddsmakers in Las Vegas gave the Montreal Canadiens 5:3 odds against winning the Stanley Cup. What probability did they assign for the Canadiens to win the Stanley Cup?

6. Calculate the expected sum of the 2 dice from Example 5.

CHAPTER NINE

Probability with permutations and combinations

In this chapter we will look at examples of probability where the calculations involve permutation or combination formulas. The first thing to remember is that each trial is dependent on the outcome of previous trials (a trial's calculation is affected by previous trials) because replacement is **not** permitted.

EXAMPLE 1

Five boys and 7 girls have signed up for a ski trip. Only 4 will be chosen at random to go on the trip. Determine the probability that:

(a) all will be boys.
(b) all will be girls.
(c) there will be 2 boys and 2 girls.

It is a good idea to calculate the denominator first because there are no restrictions.
Here, we would find the number of ways of selecting 4 people from 12.
Order is not important, so we use combinations.

Denominator: n(S) = C(12,4) = 495

(a) $P = \dfrac{C(5,4)}{C(12,4)}$ *To calculate the numerator, we select the 4 boys from the original 5.*

    $= \dfrac{5}{495}$

    $= \dfrac{1}{99}$

(b) $P = \dfrac{C(7,4)}{C(12,4)}$ *To calculate the numerator, we select the 4 girls from the original 7.*

    $= \dfrac{35}{495}$

    $= \dfrac{7}{99}$

(c) $P = \dfrac{C(5,2) \times C(7,2)}{C(12,4)}$ *To calculate the numerator, we select 2 boys and 2 girls from the original 5 boys and 7 girls.*

    $= \dfrac{210}{495}$ *Notice that the denominator has not changed even though the*

    $= \dfrac{14}{33}$ *numerator separates the boys and the girls.*

EXAMPLE 2

In a swim meet, there are 8 entries, 3 of whom come from the Halifax Swim Club. If we assume that their abilities are about the same, what is the probability that:

(a) the Halifax swimmers, Anna, Beth and Candice, will finish first, second and third, respectively?

(b) there will be no Halifax swimmers in the top 3?

Again, calculate the denominator first: n(S) = P(8,3) = 336. We use permutations here because order is important.

(a) $P = \dfrac{P(3,3)}{P(8,3)}$ *To calculate the numerator, we arrange all 3 of the 3 Halifax swimmers.*

$= \dfrac{6}{336}$

$= \dfrac{1}{56}$

(b) $P = \dfrac{P(5,3)}{P(8,3)}$ *To calculate the numerator, we arrange 3 of the 5 non-Halifax swimmers.*

$= \dfrac{60}{336}$

$= \dfrac{5}{28}$

EXAMPLE 3

In Lotto 6/49, 6 different numbers must be selected from the numbers 1 to 49. Calculate the odds against winning first prize.

$p = \dfrac{1}{C(49,6)} = \dfrac{1}{13\ 983\ 816}$ *To calculate the denominator, select 6 numbers from the original 49 without regard to order.*

$q = 1 - p$ *To calculate the numerator, there is only 1 winning set of numbers.*

$= \dfrac{13\ 983\ 815}{13\ 983\ 816}$

Odds against = 13 983 815 : 1

Hypergeometric Distribution

In a Hypergeometric Distribution, the repeated trials are **dependent** upon the outcomes of previous trials. As the trials continue, **replacement is not permitted**. The events are **not equally likely**.

$$P = \frac{C(a,x) \times C(b,r-x)}{C(n,r)}$$

where **n** is the number of elements available, **r** is the number of trials in which **x** elements are successfully selected from **a** available elements and the rest, **r - x**, are selected from the remaining **b** elements. *Note: **a + b = n***

EXAMPLE 4

Show the complete probability distribution for the number of hearts in a 5-card hand dealt from a deck of 52 cards. Determine the expected number of hearts.

Denominator first: C(52,5) = 2 598 960
For the numerator, there are 13 hearts, so a = 13.
There are 39 other cards, so b = 39.

NUMBER OF HEARTS x	NUMERATOR P(x)	PROBABILITY	xP(x)
0	C(13,0) x C(39,5)	0.22153	0
1	C(13,1) x C(39,4)	0.41142	0.41142
2	C(13,2) x C(39,3)	0.27428	0.54856
3	C(13,3) x C(39,2)	0.08154	0.24462
4	C(13,4) x C(39,1)	0.01073	0.04292
5	C(13,5) x C(39,0)	0.000495	0.002475
SUM			1.249995

$$\sum x\mathrm{P}(x) = 1.25$$

∴ *On the average, we would expect to be dealt 1.25 hearts.*

A short way of calculating the **expectation** for the **hypergeometric distribution** is:

$$\mathrm{E}(x) = \frac{ra}{n}$$

Applying the formula to Example 4:

$$\mathrm{E}(x) = \frac{5 \times 13}{52}$$

$$= 1.25$$

Again, the expectation is 1.25 hearts.

EXAMPLE 5

Of 100 circuit boards produced in a production run at a factory, 8 are defective. If a random sample of 5 is taken:
(a)   what is the probability that at least 1 is defective?
(b)   what is the expected number of defective circuit boards?

$$n = 100, \quad a = 8, \quad r = 5$$

(a) $\mathrm{P} = 1 - \dfrac{\mathrm{C}(8,0) \times \mathrm{C}(92,5)}{\mathrm{C}(100,5)}$ *It is easier to take 1 minus*
            the probability of none
  $= 0.3468$ *defective.*

∴ x = 0.

(b) $\mathrm{E}(x) = \dfrac{5 \times 8}{100}$ ∴ *On average, there will be*
            0.4 defective circuit boards.
  $= 0.4$

EXAMPLE 6

A wildlife group caught and tagged 80 bears. A month later, 20 bears were caught and 6 of them had been tagged. Estimate the bear population in that area.

If we allow the outcome of 6 to be the expected outcome, E = 6, then r = 20 (20 selected) and a = 80 (80 tagged bears available):

$$6 = \frac{20 \times 80}{n}$$

$$n = 266.67$$

∴ *There are about 267 bears in the area.*

Study solving strategies for Hypergeometric Distribution

1. Determine if repetition (replacement) is permitted. If not, the methods in this chapter are appropriate.
2. Determine if order is appropriate
3. Calculate the denominator, n(S), first.
4. The denominator includes all arrangements or combinations — without the specific restriction stated by the problem. Errors frequently occur here.
5. The numerator is where the grouping of elements occurs.
6. Assign values to n, a, b, x, and r first, before solving.
7. Remember that $a + b = n$. ALWAYS!!
8. When calculating expectation, be careful to assign correct values to r (number selected), a (number available of the stated type) and n (total number available in the trial).
9. In a multiple-part question, generally try to ignore the descriptions and answers from preceding parts. They are usually not connected. This is an easy mistake to make.

PRACTICE EXERCISE 9

1. A box contains 6 red, 3 green, 4 brown, 5 purple and 2 yellow candies. Four candies are selected at random.
    (a) What is the probability that 2 are red?
    (b) What is the probability that 3 are green and 1 is yellow?
    (c) What is the expected number of purple candies?

2. When buying a lottery ticket, a person must select 4 different numbers between 1 and 20 inclusive.
    Show the complete probability distribution of outcomes for the number of perfect squares that are on the winning ticket. Calculate the expected number of perfect squares, using both methods.

3. A group contains 12 women and 15 men. Seven are selected at random.
    (a) What is the probability that at least 3 are women?
    (b) What is the probability that all are men?
    (c) What is the expected number of men?

4. A 13-card hand was dealt from a deck of 52 cards. What is the probability that
    (a) there are no spades?
    (b) there are 2 hearts, 3 diamonds, 5 spades and 3 clubs?
    (c) there are 3 or fewer face cards (J, Q, K)?

5. A computer randomly selects 4 numbers between 1 and 10, inclusive. What is the expected number of:
    (a) even numbers?
    (b) multiples of 3?

6. In a study of Canada geese, 200 geese of a known population of 1200 were caught and tagged. Later, 50 geese were caught. How many would be expected to have been tagged?

7. Ten slips of paper, each containing the name of a different child, were placed into a hat. Two slips were drawn at random and the names recorded in the order drawn.
    What are the odds against Barb's and Alice's names being drawn in that order?

CHAPTER TEN

Independent events

Independent events are given that name because the outcome of one trial has no effect on further trials.

EXAMPLE 1

A die is rolled 3 times. Determine the probability that the first roll shows a 4, the second shows a 3 and the third shows a 6.

There are 6 faces on a standard die.
$\therefore$ *Denominator = 6 x 6 x 6 = 216*

$$P = \frac{1 \times 1 \times 1}{6 \times 6 \times 6} \quad or \quad \frac{1}{6} \times \frac{1}{6} \times \frac{1}{6}$$

$$= \frac{1}{216}$$

Because each roll of the die has no effect on the other rolls, each roll is independent. As a result, we simply multiply their individual probabilities.

Events A and B are **independent** if
and only if **P(A and B) = P(A) x P(B)**

This can also be written as
P(A $\cap$ B) = P(A) x P(B).

EXAMPLE 2

On a Las Vegas roulette wheel, there are 18 black, 18 red and 2 green numbers. Determine the probability that on 4 consecutive spins, the results are:

(a)   {black, black, green, red}
(b)   {red, black, red, black}

Each trial (spin) is independent.
There are 38 numbers from which to choose, over 4 trials.

(a) $P = \dfrac{18}{38} \times \dfrac{18}{38} \times \dfrac{2}{38} \times \dfrac{18}{38}$

        $= 0.0056$

(b) $P = \left(\dfrac{18}{38} \right)^4$

        $= 0.0503$

The solutions change when the word **and** is changed to the word **or**. Generally, we multiply when "and" is used and add when given different cases connected with the word "or", but, as the following example shows, we must consider the intersecting possibilities.

EXAMPLE 3

If a roulette wheel is spun **twice**, what is the probability that a green number will come up first or a red number will come up second?

If we simply calculate this as
P(green, any color) + P(any color, red)
we would do the following:

$$P = \frac{2}{38} \times \frac{38}{38} \times \frac{38}{38} \times \frac{18}{38}$$

$$= \frac{10}{19}$$

However, it is more complex than that. The following Venn diagram will help explain:

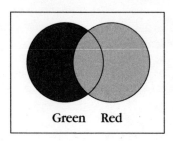

Green   Red

The circles represent the green and red numbers. The centre section represents the intersection of the two sets, where the combination {green, red} has now been included in our calculations twice. Therefore, we need to subtract P(green, red) once.

$P = P(green,\ any\ color) + P(any\ color,\ red) - P(green, red)$

$$= \frac{2}{38} \times \frac{38}{38} + \frac{38}{38} \times \frac{18}{38} - \frac{2}{38} \times \frac{18}{38}$$

$$= \frac{181}{361}$$

"or" Events

The probability of events A or B can be calculated as:

$$P(A \cup B) = P(A) + P(B) - P(A \cap B)$$

EXAMPLE 4

In a family of 4 children, determine the following probabilities:

(a) That the first is a girl **and** the next 3 are boys.

(b) That the first is a girl **or** the next 3 are boys.

Each sex has an independent probability of 1/2:

(a) $P = P(girl) \times P(3 \ boys)$

$$P = \frac{1}{2} \times \left(\frac{1}{2}\right)^3$$

$$= \frac{1}{16}$$

(b) $P = P(girl) + P(3 \ boys) - P(girl \cap 3 \ boys)$

$$P = \frac{1}{2} + \left(\frac{1}{2}\right)^3 - \frac{1}{2} \times \left(\frac{1}{2}\right)^3$$

$$= \frac{11}{16}$$

The following example is informally known as the "**Birthday Problem**." It is fascinating because of its surprising results.

EXAMPLE 5

There are 25 people in a room. What is the probability that at least 2 of them have the same birthday?

The easy way of solving this problem is to take
1 - P(all different).
Each person has a choice of 365 days for his/her
birthday.
Each is an independent choice so the denominator
is 365^{25}.
The numerator, however, is not independent. If each
person is to have a different birthday, we must use
permutations or P(365,25).

$$P = 1 - \frac{P(365,25)}{365^{25}}$$

$$= 0.5687$$

Note the high probability. In fact, it first reaches 0.5 with 23 people. Thus, there is a better than 50% chance for 2 people in your class to have the same birthday.

EXAMPLE 6

An apartment tower contains 30 floors, each with the same number of residences. At a meeting, 14 residences were represented. What is the probability that at least 2 residences were from the same floor?

We use the "Birthday Problem" method here.

$$P = 1 - P(\textit{none the same})$$

$$= 1 - \frac{P(30,14)}{30^{14}}$$

$$= 0.9735$$

Conditional Probability

The following example illustrates **Conditional Probability**, so-called because of a **given condition**.

***EXAMPLE* 7**

When two friends, Brian and Hugh, play ping-pong, Hugh wins 60% of the time and Brian 40% of the time. However, Hugh will win 2 in a row 50% of the time. If it is given that Hugh has won the first game of a match, what is the probability that he wins the second?

> *Because there is "given" information that Hugh has won the first game, we only need to consider the 60% of the games that match the "given" information. We do not need to consider Brian's 40%.*
>
> *As a result, we allow the denominator to be 60%.*
>
> *Thus, the numerator = 50% representing Hugh's double win.*

$$P = 1 - \frac{P(\text{Hugh} \cap \text{Hugh})}{P(\text{Hugh})}$$

$$= \frac{50\%}{60\%}$$

$$= 0.8333$$

$$= 83.33\%$$

Using the above example as a guide, the sample space, **n(S),** can be replaced by the probability of the **given condition** which, in the box below, is **P(B)**.

Conditional Probability

The probability of event A, given event B, can be found:

$$P(A|B) = \frac{P(A \cap B)}{P(B)}$$

EXAMPLE 8

A recent national survey indicated that 15% of the population is left-handed, 10% excel at mathematics and 3% are left-handed and excel at mathematics. Given that a certain individual is left handed, what is the probability of that person also excelling at mathematics?

Again, we need only consider the 15% of the population that is left-handed.

$$P = \frac{P(Math \cap Left)}{P(Left)}$$

$$= \frac{3\%}{15\%}$$

$$= 0.2$$

$$= 20\%$$

EXAMPLE 9

In a factory, 45% of the output is from Machine A, 2% of which is defective, and 55% is from Machine B, 1.5% of which is defective. Determine the probability that a certain item that is defective was manufactured on Machine B.

$$P(B/Defective) = \frac{P(B \cap Defective)}{P(Defective\ from\ A\ or\ B)}$$

$$= \frac{P(B) \times P(Defective)}{P(A) \times P(Defective) + P(B) \times P(Defective)}$$

$$= \frac{0.55 \times 0.015}{0.45 \times 0.02 + 0.55 \times 0.015}$$

$$= 0.4783$$

$$= 47.83\%$$

64

PRACTICE EXERCISE 10

1. If P(A) = 0.3, P(B) = 0.24 and P(A ∩ B) = 0.06, are events A and B independent? Why?

2. A card is drawn from a deck 5 times and replaced each time. Determine the probability that
   (a) the first 2 cards are kings and all 5 are red.
   (b) the first 2 cards are kings or all 5 are red.

3. Two dice were rolled and the result is known to be even. What is the probability that the sum is 8?

4. What is the probability that, in a group of 10 students, at least 2 will have the same
   (a) birthday?
   (b) birth month?

5. In a local school, 48% of the students are male and 52% are female. Portable radios are used by 15% of the males and by 24% of the females. A portable radio was turned into the lost and found. What is the probability that it is owned by a female student?

6. Eight people were each asked to close their eyes and point to 1 of 50 numbers on a board. What is the probability that at least 2 of the people selected the same number?

7. A bag contains 8 red, 5 green and 7 blue balls. A ball is removed and returned 3 times.
   Determine the following probabilities:
   (a) The first is green and the second and third are red.
   (b) The first is blue, given that it is known not to be red.
   (c) The first two are red or the third is blue.
   (d) The first two are green or all three are red.

CHAPTER ELEVEN

Binomial Distribution

The **binomial probability distribution** is so called because it resembles the **binomial expansion**. (See Chapter 7)

Recall the following binomial expansions from Chapter 7:

$$(x + y)^3 = C(3,0)x^3 + C(3,1)x^2y^1 + C(3,2)x^1y^2 + C(3,3)y^3$$

$$\left(\frac{1}{6} + \frac{5}{6}\right)^3 = C(3,0)\left(\frac{1}{6}\right)^3 + C(3,1)\left(\frac{1}{6}\right)^2\left(\frac{5}{6}\right)^1 + C(3,2)\left(\frac{1}{6}\right)^3\left(\frac{5}{6}\right)^3 + C(3,3)\left(\frac{5}{6}\right)^3$$

EXAMPLE 1

A single die is rolled a total of 3 times and the number of five's is recorded. What is the probability that 2 five's show?

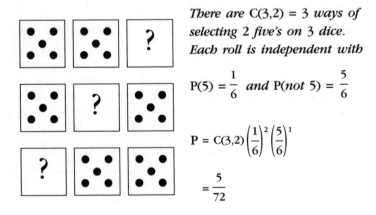

There are C(3,2) = 3 ways of selecting 2 five's on 3 dice. Each roll is independent with

$$P(5) = \frac{1}{6} \ \ and \ P(not \ 5) = \frac{5}{6}$$

$$P = C(3,2)\left(\frac{1}{6}\right)^2\left(\frac{5}{6}\right)^1$$

$$= \frac{5}{72}$$

Notice the similarities with the binomial expansions above.

The Binomial Distribution

The Binomial Distribution is a distribution of **counting** the number of **successful** independent outcomes. In an event containing **n** trials, the probability of **r** successful trials will be:

$$C(n,r)p^r q^{n-r}$$

where p is the probability of the event being successful on a single trial and q is the probability of it being **unsuccessful** on a **single trial**. Note: $p + q = 1$.

EXAMPLE 2

In a series of experiments, a mouse is placed into a maze. In 75% of the experiments, the mouse reaches the food at the end of the maze.

(a) Show the complete probability distribution of the number of times it reaches the food over a series of 5 experiments.

(b) Calculate the expectation.

NUMBER OF SUCCESSES $(x = r)$	PROBABILITY $C(5,r)(0.75)^r(0.25)^{n-r}$	$x P(x)$
0	0.00098	0
1	0.01465	0.01465
2	0.08789	0.17578
3	0.26367	0.79102
4	0.39551	1.58203
5	0.2373	1.18652
SUM		3.75

$$\sum x P(x) = 3.75$$

∴ *On average, the mouse will be successful in finding the food 3.75 times.*

A short way of calculating the expectation for the Binomial Distribution is:

$$E(x) = n \times p$$

Applying the formula to Example 2:

$$E(x) = 5 \times (0.75) = 3.75$$

∴ *Again, the expectation is 3.75.*

EXAMPLE 3

In a multiple-choice quiz, there are 6 questions. Each question has 5 possible answers. If a student guesses at each question:
(a) what is the probability of getting 4 answers correct?
(b) what is the probability of passing?
(c) what is the expected number of correct answers?

Each question has 5 possible answers and each question is independent.

$$\therefore p = \frac{1}{5} \quad and \quad q = \frac{4}{5}$$

There are 6 questions, so n = 6.

(a) *Here, there are 4 successes, so r = 4.*

$$P = C(6,4)\left(\frac{1}{5}\right)^4\left(\frac{4}{5}\right)^2$$

$$= 0.01536$$

(b) *Passing requires 3, 4, 5 or 6 correct answers, so we must add those four cases.*

$$P = C(6,4)\left(\frac{1}{5}\right)^3\left(\frac{4}{5}\right)^3 + C(3,1)\left(\frac{1}{5}\right)^4\left(\frac{4}{5}\right)^2 + C(6,5)\left(\frac{1}{5}\right)^5\left(\frac{4}{5}\right)^1 + C(6,6)\left(\frac{1}{5}\right)^6$$

$$= 0.09888$$

(c) $$E(x) = n \times p$$

$$= 6 \times \frac{1}{5}$$

$$= 1.2$$

On average, there will be 1.2 correct answers.

EXAMPLE 4

In a factory, 4% of the widgets manufactured are defective. What is the probability that at least one of a batch of 50 widgets are defective?

For "at least one," it is best to try the approach of 1 - P(none defective).

$$P = 1 - C(50,0)(0.04)^0(0.96)^{50}$$

$$= 0.8701$$

Study strategies for the Binomial Distribution

1. Each trial must be independent.
2. The outcomes can be described in only two ways: success or failure.
3. The problem discusses counting the number of successful (or unsuccessful) outcomes.
4. The number of trials is specified.
5. Do not confuse the formula for expectation with those for other distributions.
6. Do not confuse this distribution with the hypergeometric distribution. Hypergeometric trials are dependent.

PRACTICE EXERCISE 11

1. Two dice are rolled a total of 6 times.
   (a) Show the complete probability distribution of outcomes for the number of times there is a sum of 5 on the 2 dice.
   (b) Calculate the expectation using both methods.

2. A baseball player has a batting average of 0.325. In a given game, what is the probability that after 5 times at bat, the player will have had at least 2 hits?

3. A certain type of rocket has a chance of failure of 1 in 40.
   (a) What is the probability that, in 100 launches of the rocket, fewer than 3 will fail?
   (b) What is the expected number of failures in 100 launches of the rocket?

4. Mr. Tam is a math teacher. He calculated that the probability of having perfect attendance in his math class on any given day is 0.1. During a 10-day period:
   (a) what is the probability of Mr. Tom having perfect attendance on 6 days?
   (b) what is the probability of Mr. Tom having perfect attendance on only 2 days?
   (c) what is the expected number of days with perfect attendance in Mr. Tom's class?

5. A game involves cutting a deck of cards a total of 15 times, looking at the card each time, then replacing the card in the deck.
   (a) What is the probability of cutting four aces?
   (b) What are the odds against cutting 5 spades?
   (c) What is the probability of cutting at least 13 red cards?
   (d) What is the expected number of hearts?
   (e) What is the expected number of kings?

6. A lottery ticket has 5 digits between 1 and 8. Susan's lucky number is 7.
   (a) What is the probability that there are 2 sevens on the winning ticket?

(b) What is the probability that there are more than 2 sevens on the winning ticket?

(c) What is the expected number of sevens on the winning ticket?

7. In a group of 30 people, what is the probability that at least 2 people have the same birthday as you?

8. A gambling game consists of rolling 4 dice with a payout of $2 for each 6 rolled.

(a) What is the expected payout for each game?

(b) If it costs $2 to play, is this a fair game (one where the average payout equals the cost to play)?

CHAPTER TWELVE

Geometric Distribution

The Geometric Probability Distribution is one in which we find the probability of an event occurring **successfully after a series of failures.** The first example illustrates this concept.

EXAMPLE 1

In a dice game, a person can only win by rolling a sum of 7. If a person tries a number of times before rolling a seven, what is the probability that it will occur on the fourth roll?

In an example such as this, there will be 3 failures followed by a single successful trial.

$$p = P(winning) = \frac{6}{36} = \frac{1}{6} \quad \therefore \ q = P(failure) = \frac{5}{6}$$

$$P = \left(\frac{5}{6}\right)^3 \left(\frac{1}{6}\right)$$

$$= \frac{125}{1296}$$

Geometric Distribution

The probability of success after x unsuccessful independent trials is:

$$P(x) = q^x p$$

where p is the probability of **success** and q is the probability of **failure** on an individual trial. x is known as the **waiting time** before success.

EXAMPLE 2

A particular hockey player scores on 18% of his shots. What is the probability that he will first score in fewer than 3 shots?

$$\text{Given that } p = 0.18$$
$$\therefore \ q = 0.82$$

$$
\begin{aligned}
\text{P}(x<2) &= \text{P}(0) + \text{P}(1) \\
&= (0.82)^0(0.18) + (0.82)^1(0.18) \\
&= 0.3276
\end{aligned}
$$

Remember x is the "waiting time," so x < 2 not <3.

EXAMPLE 3

Show the development of the geometric probability distribution for cutting a deck of cards until a heart shows. Calculate the expectation.

$$p = \frac{1}{4} = 0.25 \qquad q = \frac{3}{4} = 0.75 \qquad \text{P}(x) = (0.75)^x(0.25)$$

WAITING TIME x	PROBABILITY $\text{P}(x)$	$x\text{P}(x)$
0	0.25	0
1	0.1875	0.1875
2	0.140625	0.28125
3	0.105469	0.3164
4	0.079102	0.3164
5	0.059326	0.2966

Because of the nature of the geometric distribution, this table of values could be infinitely long. To find the expectation, we do not use $\sum x\text{P}(x)$. We use the formula below, which can be proven using the infinite geometric series formula.

To find the expected **waiting time** before success, use the formula:

$$E(x) = \frac{q}{p}$$

Returning to Example 3:

$$E(x) = \frac{q}{p}$$

$$= \frac{0.75}{0.25}$$

$$= 3$$

The expected waiting time before a heart is 3 cuts.
This means that, on average, there will be 3 failures followed by a heart.

EXAMPLE 4

If you were to randomly select people until you found one whose birthday is on a Sunday:

(a) what is the probability that you would not be successful until the 5th person?

(b) what is the probability of being successful within the first 3 people?

(c) what is the expected waiting time before success?

$$p = \frac{1}{7} \quad q = \frac{6}{7}$$

(a) *"Until the fifth person" means 4 failures initially.*

$$P(4) = \left(\frac{6}{7}\right)^4 \left(\frac{1}{7}\right)$$

$$= 0.07711$$

(b) *"Within the first 3" means up to 2 failures initially.*

$$P(< 3) = P(0) + P(1) + P(2)$$

$$= \left(\frac{1}{7}\right) + \left(\frac{6}{7}\right)^1 \left(\frac{1}{7}\right) + \left(\frac{6}{7}\right)^2 \left(\frac{1}{7}\right)$$

$$= 0.37026$$

(c)
$$E(x) = \frac{q}{p}$$

$$= \frac{\frac{6}{7}}{\frac{1}{7}}$$

$$= 6$$

On average, there will be 6 failures before finding a person whose birthday is on a Sunday.

Study strategies for the Geometric Distribution

1. Each trial must be independent.
2. Look for clue words, such as "before" or "until" an event occurs.
3. Remember "waiting time" means "number of failures" before success.
4. Do not confuse the expectation formula with those of other distributions.
5. Do not get confused with the binomial distribution, which counts the total number of successes in n trials.

PRACTICE EXERCISE 12

1. A box contains 4 red and 8 green balls. A ball is selected at random and replaced continuously until a red ball is selected.
   (a) What is the probability that the first red ball will be on the seventh draw?
   (b) What are the odds against being successful within the first 4 draws?
   (c) What is the expected waiting time before selecting a red ball?

2. A new drug, Mitocol, is reportedly successful as a headache remedy 65% of the time. What is the probability that, in a random sample of people with headaches:
   (a) Mitocol is not successful until the fourth person?
   (b) Mitocol is not successful until the fifth person?

3. A student has a summer job selling replacement windows by telephone. It is known that 9 out of 10 people hang up before the student can even give a sales pitch.
   (a) This student was not able to make a sales pitch until the fifteenth call. What is the probability of this occurring?
   (b) What is the expected number of calls before being able to make a sales pitch?

4. At a child's birthday party, the children are playing a game involving a spinner that points to 1 of 5 children. What is the probability that the spinner will first point to little Johnny within the first 3 spins?

5. It is estimated that 74% of a store's customers use a credit card and the rest pay by cash.
   (a) What is the probability that the sixth customer is the first to pay by cash?
   (b) What is the probability that the first to pay by cash will be within the first 5 customers?
   (c) What is the expected waiting time until a customer pays by cash?

APPENDICES

Sample examination 1

INSTRUCTIONS:
1.   Provide complete solutions for each of the following questions.
2.   Calculators are permitted.
3.   Time allotted: 2 hours

---

1.   State whether each of the following is True or False
     (a) 0! = 1   (b) P(n,r) = P($n,n - r$)

     (c) C(72,57) = C(72,15)   (d) -4! = (-4)(-3)(-2)(-1)

     (e) C(15,0) + C(15,1) + C(15,2) + ... + C(15,15) = 2^{15}

2.   A lottery ticket has 5 different numbers between 1 and 15, followed by 2 letters. The order of the numbers and letters is important. How many different lottery tickets can be printed?

3.   In how many ways could 10 children be arranged:
     (a) in a row?   (b) in a circle?

4.   There are 5 history, 3 math and 2 French books on a shelf.
     (a) In how many ways could they be arranged on the shelf?
     (b) In how many ways could they be arranged, if the French books are to be kept apart?

5. Four girls and 6 boys are to be lined up in a row.
   (a) In how many ways could this be done?
   (b) If the boys and girls must be grouped separately, in how many ways could this be done?

6. How many 5-digit numbers are there that do not contain the digits 5 or 9?

7. Solve algebraically for $n \in W$.

$$P(n + 2,3) = 2(n - 1)P(n,2)$$

8. How many subsets of 5 elements can be formed from a set containing 11 elements?

9. In how many ways could 15 different cookies be divided evenly among 3 children?

10. In how many ways could 12 identical toys be placed into 4 distinguishable toy boxes, if each box must contain at least 1 toy?

11. There are 1 red, 1 green, 1 blue, 3 white, 4 orange and 5 purple balls in a bag. If a person reaches into the bag, in how many ways could that person take out at least 1 ball?

12. There are 10 points (no 3 in a straight line) drawn on a page.
    (a) How many triangles could be drawn, using these points as vertices?
    (b) How many different polygons could be formed?

13. On the game board shown below, a player may move 1 space at a time, diagonally left or diagonally right. If an X is encountered, no move may take place into that square, but the playing piece may jump over it. For the playing piece shown with a ●, how many paths are there to the top of the game board?

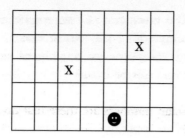

14. Use the binomial theorem to expand and simplify each of the following.

    (a) $(3x + 4)^6$

    (b) $\left(m^2 + \dfrac{2}{m}\right)^4$

15. Write in the form $(a + b)^n$.

    $x^{15} - 20x^{12} + 160x^9 - 640x^6 + 1280x^3 - 1024$

16. Two dice are rolled. Determine the probability that:
    (a) both show the same number.
    (b) the sum is divisible by 4.
    (c) the sum is divisible by 3 or by 4.

17. There are 26 letters on a keypad. Four keys were touched at random. What is the probability that:
    (a) at least 2 letters are the same?
    (b) the letters are in ascending order?
    (c) the letter "A" was touched at least once?

18. During a particular season, 45% of the Vancouver Canuck's goals were in away games and 55% were in home games. Trevor Linden scored 12% of the away goals and 15% of the home goals. Given that Trevor Linden scored a goal in a particular game, what is the probability that it was in an away game?

19. In a jar of 100 green and yellow jelly beans, 20 are yellow. A random sample of 12 jelly beans is taken.
(a) How many should be expected to be yellow?
(b) What is the probability that 7 are yellow?

20. (a) What are the odds against a person drawing 2 face cards in 5 tries (with replacement) from a standard deck of 52 cards?
(b) Calculate the expectation.

21. Cars driving along a certain street have a 1 in 3 chance of encountering a red light at each intersection.
(a) What is the probability that the first red light encountered by a particular car does not occur until the fifth intersection?
(b) What is the expected waiting time until the first red light is encountered?

Sample examination 2

1. Evaluate each of the following.
   (a) 10!   (b) P(15,6)   (c) C(9,7)
   (d) C(11,0) + C(11,1) + C(11,2) + … + C(11,11)

2. A standard die is rolled 5 times. How many outcomes are possible?

3. In how many ways could 8 different cards be arranged
   (a) in a row?
   (b) in a circle?

4. A landscaper has 3 cedar, 5 maple, 2 pine and 2 oak trees to plant along a fence. In how many ways could this be done?

5. In how many ways could 8 cars be arranged in a circle, if the Lotus and Ferrari must not be directly opposite each other?

6.　In how many ways could the letters in the word MINIMUM be arranged if:
(a) the arrangement must begin with an I?
(b) the U must not come before the I's?
(c) the N and U must be together?

7.　In how many ways could a selection of 7 books be chosen from a shelf containing 13 books?

8.　Twelve different prizes are to be distributed evenly amongst 4 people. In how many ways could this be done?

9.　How many factors are there of the number 4158?

10.　There are 4 red, 5 green and 3 blue balls in a bag. In how many ways could at least 1 ball be selected from the bag?

11.　Ten points are drawn as shown. In how many ways could a triangle be formed, using these points as vertices?

12.　Solve algebraically for $n \in N$.
$$C(n,2) = 2C(n - 2,2) - 24$$

13.　Use the binomial theorem to expand and simplify each of the following.

(a)　$(4a + b)^5$

(b)　$\left(y^2 - \dfrac{2}{y^2}\right)^6$

14.　Rewrite the following in the form $(a + b)^n$.
$x^8 - 20x^6y + 150x^4y^2 - 500x^2y^3 + 625y^4$

15. Sticks are placed on top of each other so that each stick intersects with each other stick exactly once, as shown below. Use patterns in Pascal's Triangle to determine a formula for the total number of intersections for n sticks. Show the development of your formula.

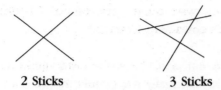

2 Sticks   3 Sticks

16. A box contains 4 red cards and 3 blue cards. Three cards are drawn at random. What is the probability that 2 tickets are blue:
    (a) if the cards are replaced after each draw?
    (b) if the cards are not replaced after each draw?

17. The odds against a golfer scoring a hole-in-one on a particular hole are 2000:1. What is the probability of:
    (a) scoring a hole-in-one?
    (b) scoring a hole-in-one on 2 successive tries?

18. There are 5 people in a room. What are the odds against at least 2 of them having been born in the same month?

19. An experiment consists of rolling 2 dice and recording the upper faces. Determine the probability that:
    (a) the sum is a prime number.
    (b) both dice show a number divisible by 3.
    (c) the sum is a prime number or it is even.

20. In establishing its criteria for admission to its degree program, Northern University makes use of its Northern Admission Test. If a person has a B average in high school, the probability of passing the test is 0.75. Otherwise, the probability is 0.48. It is known that 60% of the applicants had a B average and 40% did not. What is the probability that a person who passed the test had a B average?

21. It has been predicted that the North Shore Moose will win only 12 of their 80 games this season.

    (a) When should the fans expect the Moose to win their first game?

    (b) What is the probability that the Moose will not win until their fifth game of the season?

22. Seven girls and 5 boys sign up for a contest. Six are chosen at random.

    (a) What is the probability of selecting 2 boys?

    (b) Calculate the expected number of boys.

23. A particular basketball player is successful on 80% of her free throws. She will win $100 for each successful free throw. She will have 15 free throw attempts.

    (a) What is the probability that she will win exactly $1000?

    (b) What are her expected winnings?

Solutions to practice exercises

PRACTICE EXERCISE 1

1. (a)

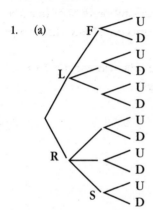

(b) $2 \times 3 \times 2 = 12$

2. $52^4 = 7\ 311\ 616$

3. $2^6 = 64$

4. $3^3 = 27$

5. $10^5 44^1 = 25\ 600\ 000$

PRACTICE EXERCISE 2

1. (a) 210   (b) 95 040
2. (a) $(n + 2)!$
   (b) $n(n - 1)(n - 2)$
   (c) $(n-2)(n-3)(n-4)$
   (d) $n(n+1)$

PRACTICE EXERCISE 2 (cont'd)

3. (a) $12! = 479\ 001\ 600$
   (b) $1 \times 11! = 39\ 916\ 800$
4. (a) $P(15,5)15 = 5\ 405\ 400$
   (b) $P(15,6) = 3\ 603\ 600$
5. (a) $P(7,4) = 840$
   (b) $7^4 = 2401$
   (c) $3P(6,3) = 360$
6. $3! \times 5! = 720$

PRACTICE EXERCISE 3

1. (a) $\dfrac{7!}{2!3!} = 420$

   (b) $\dfrac{11!}{2!2!2!2!} = 2\ 494\ 800$

   (c) $\dfrac{8!}{2!2!2!} = 5040$

2. $1 \times \dfrac{7!}{2!2!} = 1260$

3. $\dfrac{6!}{3!} = 120$

4. $\dfrac{7!}{2!2!2!} = 630$

5. $\dfrac{2 \times 5!}{2!2!} = 60$

6. $\dfrac{12!}{3!4!3!} = 554\ 400$

7. $\dfrac{12!}{4!4!4!} = 34\ 650$

8. $\dfrac{9!}{4!5!} = 126$

9. $\dfrac{10!}{3!2!5!} = 2520$

PRACTICE EXERCISE 4

1. $C(15,10) = 3003$
2. (a) $C(26,13)$
   (b) $C(13,2) \times C(13,3) \times C(13,5)$
      $\times C(13,3)$
   (c) $C(40,13)$
3. $C(15,5) \times C(10,5) \times C(5,5) = 756$
4. $C(17,2) = 136$
5. (a) $C(10,2) = 45$
   (b) $C(10,3) = 120$
   (c) $C(10,2) \times C(8,3) = 2520$
6. (a) 495   (b) 1   (c) 1   (d) 1
   (e) 2520   (f) 34 650
7. $n = 6$

PRACTICE EXERCISE 5

1. (a) $2^7 = 128$
   (b) $2^{10} = 1024$
   (c) $4 \times 3 \times 2^4 = 192$
2. $2^7 - 1 = 127$
3. $51\ 975 = 3 \times 3 \times 3 \times 5 \times 5 \times 7 \times 11$
   $4 \times 3 \times 2^2 - 2 = 46$
4. $2^n = 8192$
   $n = 13$
5. $5 \times 3 \times 4 \times 6 - 1 = 359$
6. (a) $2^8 = 256$
   (b) $2^{16} = 65\ 536$
   (c) $2^{32} = 4\ 294\ 967\ 296$

PRACTICE EXERCISE 6

1. $\dfrac{9!}{2!4!} = 7560$
2. $1 \times 8! = 40\ 320$
3. Reversing the necklace does not
   change it $\Rightarrow$ divide by 2.
   $8! \div 2 = 20\ 160$
4. $C(6,2) - 6 = 9$
5. $P(16,2) \times C(14,3) = 87\ 360$
6. $1 \times 9! - 1 \times 8! \times 2! = 282\ 240$
7. Place a divider into 3 of the 19
   spaces between the 20 books $\Rightarrow$ 4
   groups.
   $C(19,3) = 969$
8. (a) $40 \times 1^4 - 4 \times 1^4 = 36$
   (b) $40 \times 4^4 - 40 \times 1^4 = 10\ 200$
9. (a) $C(14,5) = 2002$
   (b) $C(14,5) - C(8,0) \times C(6,5)$
      $- C(8,5) \times C(6,0) = 1940$
10. Total - both 8's, no 16
    $3 \times 2^4 - 2^4 = 32$

3		7		7	
	3		4		3
1		**X**		3	
	1		2		1
		1		1	
			0		

3 + 7 + 7 = 17

2.

n	1	2	3	4	5
Number of cirlces	1	3	6	10	15

Following the diagonal in Pascal's Triangle, we have:

C(2,2), C(3,2), C(4,2), C(5,2), C(6,2), …which can be generalized as C(n + 1,2).

3. (a) $4096x^6 - 6144x^5y + 3840x^4y^2 - 1280x^3y^3 + 240x^2y^4 - 24xy^5 + y^6$

(b) $x^8 + 12x^6y + 54x^4y^2 + 108x^2y^3 + 81y^4$

(c) $32x^5 + 80x^2 + 80x^{-1} + 40x^{-4} + 10x^{-1} + x^{-10}$

PRACTICE EXERCISE 8

1. (a) $(1/2)^2 = 1/4$
   (b) $(1/2)^2 = 1/4$
   (c) $2 \times (1/2)^2 = 1/2$

2. (a) 8:1192 = 1:149
   (b) 1:1199
   (c) 7:1193

3.

OUTCOME	PROBABILITY
Red	2/11
Green	3/11
Blue	1/11
White	5/11

4.

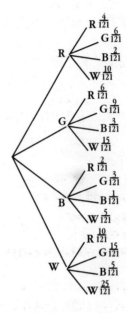

5. $q{:}p = 5{:}3$

$$\therefore p = \frac{3}{5 + 3}$$

$$= \frac{5}{8}$$

6. E(x) = $\dfrac{252}{36}$ = 7

The average sum is 7.

1. (a) $P = \dfrac{C(6,2)(C14,2)}{C(20,4)} = 0.2817$

   (b) $P = \dfrac{C(3,3)(C2,1)}{C(20,4)} = 0.0004$

   (c) $P = \dfrac{4 \times 5}{20} = 1$

2. $P(x) = C(4,x)C(16,4-x)$

No. of Perfect Squares x	$P(x)$	$xP(x)$
0	0.3756	0
1	0.4623	0.4623
2	0.1486	0.2972
3	0.0132	0.0396
4	0.0002	0.0008
Sum		0.7999

$\sum xP(x) = 0.8$

$4.4 \div 20 = 0.8$ perfect squares

3. (a) $1 - \dfrac{C(15,7) - (C12,1)C(15,6) - C(12,2)C(15,5)}{C(27,7)}$

   $= 0.7019$

   (b) $\dfrac{C(15,7)}{C(27,7)} = 0.0072$

   (c) $\dfrac{7 \times 15}{27} = 3.9$ men

4. (a) $\dfrac{C(39,13)}{C(52,13)} = 0.0128$

   (b) $\dfrac{C(13,2)C(13,3)C(13,5)C(13,3)}{C(52,13)}$

   $= 0.0129$

4. (c) $\dfrac{C(40,13) + C(12,1)C(40,12) + C(12,2)C(40,11) + C(12,3)C(40,10)}{C(52,13)}$

   $= 0.6585$

5. (a) $\dfrac{4 \times 5}{10} = 2$ evens

   (b) $\dfrac{4 \times 3}{10} = 1.2$ multiples of 3

6. $\dfrac{50 \times 200}{1200} = 8.3$ geese

7. $P = \dfrac{P(2,2)}{P(10,2)} = \dfrac{2}{90}$

   Odds against $= 88:2$

   $= 44:1$

PRACTICE EXERCISE 10

1. $P(A) \times P(B) = 0.072$

   $\neq P(A \cap B)$

   $\therefore$ not independent

2. (a) $\left(\dfrac{2}{52}\right)^2 \left(\dfrac{1}{2}\right)^3 = 0.000\ 184\ 9$

   (b) $\left(\dfrac{4}{52}\right)^2 + \left(\dfrac{1}{2}\right)^5 - \left(\dfrac{2}{52}\right)^2 \left(\dfrac{1}{2}\right)^3$

   $= 0.036\ 98$

3. $\dfrac{\frac{5}{36}}{\frac{1}{2}} = \dfrac{5}{18}$

4. (a) $1 - \dfrac{P(365,10)}{365^{10}} = 0.1169$

   (b) $1 - \dfrac{P(12,10)}{12^{10}} = 0.9961$

5. $\dfrac{0.24 \times 0.52}{0.24 \times 0.52 + 0.15 \times 0.48} = 0.6341$

6. $1 - \dfrac{P(50,8)}{50^8} = 0.4458$

7. (a) $\dfrac{5}{18} \times \left(\dfrac{8}{20}\right)^2 = 0.04$

 (b) $\dfrac{\dfrac{7}{20}}{\dfrac{12}{20}} = 0.5833$

 (c) $\left(\dfrac{8}{20}\right)^2 + \left(\dfrac{7}{20}\right) - \left(\dfrac{8}{20}\right)^2\left(\dfrac{7}{20}\right) = 0.454$

 (d) $\left(\dfrac{5}{20}\right)^2 + \left(\dfrac{8}{20}\right)^3 = 0.1265$

PRACTICE EXERCISE 11

1. (a) $P(x) = C(6,x)\left(\dfrac{1}{9}\right)^x + \left(\dfrac{8}{9}\right)^{6-x}$

No. of Sums of 5 x	$P(x)$	$xP(x)$
0	0.04933	0
1	0.37	0.37
2	0.1156	0.2312
3	0.0193	0.0578
4	0.0018	0.0072
5	0.00009	0.00045
6	0.000002	0.000012
		0.666662

$\sum xP(x) = 0.67$

 (b) $6 \times \dfrac{1}{9} = 0.67$

2. $1 - C(5,0)(0.235)^0(0.675)^5 - C(5,1)(0.325)^1(0.675)^4 = 0.5225$

3. (a) $C(100,0)\left(\dfrac{1}{40}\right)^0\left(\dfrac{39}{40}\right)^{100} +$

 $C(100,1)\left(\dfrac{1}{40}\right)^1\left(\dfrac{39}{40}\right)^{99} +$

 $C(100,2)\left(\dfrac{1}{40}\right)^2\left(\dfrac{39}{40}\right)^{98}$

 $= 0.5422$

 (b) $100 \times \dfrac{1}{40} = 1.5$ failures

4. (a) $C(10,6)(0.1)^6(0.9)^4 = 0.00014$
 (b) $C(10,2)(0.1)^2(0.9)^8 = 0.1937$
 (c) $10 \times 0.1 = 1$ day

5. (a) $C(15,4)\left(\dfrac{1}{13}\right)^4\left(\dfrac{12}{13}\right)^{11} = 0.0198$

 (b) $P = C(15,5)\left(\dfrac{1}{4}\right)^5\left(\dfrac{3}{4}\right)^{10} = 0.165$
 Odds against $= 835:165 = 167:33$

 (c) $C(15,13)\left(\dfrac{1}{2}\right)^{13}\left(\dfrac{1}{2}\right)^2 +$

 $C(15,14)\left(\dfrac{1}{2}\right)^{14}\left(\dfrac{1}{2}\right)^1 +$

 $C(15,15)\left(\dfrac{1}{40}\right)^{15} = 0.0037$

 (d) $15 \times \dfrac{1}{2} = 7.5$ hearts

 (e) $15 \times \dfrac{1}{13} = 1.15$ kings

6. (a) $C(5,2)\left(\dfrac{1}{8}\right)^2\left(\dfrac{7}{8}\right)^3 = 0.01047$

    (b) $C(5,3)\left(\dfrac{1}{8}\right)^3\left(\dfrac{7}{8}\right)^2 +$

    $C(5,4)\left(\dfrac{1}{8}\right)^4\left(\dfrac{7}{8}\right)^1 +$

    $C(5,5)\left(\dfrac{1}{8}\right)^5 = 0.0161$

    (c) $5 \times \dfrac{1}{8} = 0.625$ sevens

7. $1 - C(30,0)\left(\dfrac{1}{365}\right)^0\left(\dfrac{364}{365}\right)^{30}$

    $- C(30,1)\left(\dfrac{1}{365}\right)^1\left(\dfrac{364}{365}\right)^{29} = 0.0031$

8. (a) $\$2 \times \left(4 \times \dfrac{1}{6}\right) = \$1.30$

    (b) $\$2 < 1.30$
    $\therefore$ not a fair game

PRACTICE EXERCISE 12

2. (a) $(0.35)^3(0.65) = 0.0279$
    (b) $(0.35)^4(0.65) = 0.009\ 75$

3. (a) $(0.9)^{14}(0.1) = 0.0229$

    (b) $\dfrac{0.9}{0.1} = 9$ calls

4. $\dfrac{1}{5} + \left(\dfrac{4}{5}\right)\left(\dfrac{1}{5}\right) + \left(\dfrac{4}{5}\right)^2\left(\dfrac{1}{5}\right)$

    $= \dfrac{61}{125}$

5. (a) $(0.74)^5(0.26) = 0.0577$
    (b) $0.26 + (0.74)(0.26) +$
    $(0.74)^2(0.26) + (0.74)^3(0.26)$
    $+ (0.74)^4(0.26)$
    $= 0.7781$

    (c) $\dfrac{0.74}{0.26} = 2.85$ customers

PRACTICE EXERCISE 12

1. (a) $\left(\dfrac{2}{3}\right)^6\left(\dfrac{1}{3}\right) = 0.0293$

    (b) $P = \dfrac{1}{3} + \left(\dfrac{2}{3}\right)\left(\dfrac{1}{3}\right) +$

    $\left(\dfrac{2}{3}\right)^2\left(\dfrac{1}{3}\right) + \left(\dfrac{2}{3}\right)^3\left(\dfrac{1}{3}\right)$

    $= \left(\dfrac{65}{81}\right)$

    Odds against = 16:65

    (c) $\dfrac{\dfrac{2}{3}}{\dfrac{1}{3}} = 2$ green

Solutions to sample examination 1

1.   (a) True   (b) False   (c) True   (d) False   (e) True

2. $P(15,5) \times P(26,2) = 234\ 234\ 000$
   $\therefore$ 234 234 000 tickets can be printed.

3.   (a) $10! = 3\ 628\ 800$
            They could be arranged in 3 628 800 ways.
     (b) $1 \times 9! = 362\ 880$
             They could be arranged in 362 880 ways.

4.   (a) $\dfrac{10!}{5!3!2!} = 2520$ They could be arranged in 2520 ways.

     (b)   Total - together $= \dfrac{10!}{5!3!2!} - \dfrac{9!}{5!3!}$

                                   $= 2016$ They could be arranged in 2016 ways.

5.   (a) $10! = 3\ 628\ 8000$ They could be arranged in 3 628 800 ways.
     (b)   Can start with a boy or a girl $\therefore$ multiply by 2.
             $4! \times 6! \times 2 = 34\ 560$ They could be arranged in 34 560 ways.

6. $7 \times 8^4 = 28\ 672$ There are 28 672 five-digit numbers that do not contain
                             a 5 or 9.

7. $P(n + 2,3) = 2(n - 1)P(n,2)$

$$\frac{(n + 2)!}{(n - 1)!} = 2(n - 1)\frac{n!}{(n - 2)!}$$

$(n + 2)(n + 1)n = 2(n - 1)n(n - 1)$

$(n + 2)(n + 1) = 2(n - 1)n(n - 1), \; n \neq 0$

$n^2 + 3n + 2 = 2n^2 - 4n + 2$

$n^2 - 7n = 0$

$n = 0$ or $n = 7$

$n \geq 2 \; \therefore \; n \neq 0$

$\therefore n = 7$

8. $C(11,5) = 462$ 462 subsets can be formed.

9. $C(15,5) \times C(10,5) \times C(5,5) = 756\ 756$
The cookies could be divided up in 756 756 ways.

10.   There are 11 spaces between the 12 toys when lined up. Select 3 spaces to form 4 groups.
$C(11,4) = 330$ The toys could be placed into the boxes in 330 ways.

11. $4 \times 5 \times 6 \times 2^3 - 1 = 959$ There are 959 ways to take out at least 1 ball.

12.   (a) $C(10,3) = 120$ There are 120 triangles.
(b) A polygon will have 3 or more vertices. Find total number of subsets minus 0, 1 or 2 points.
$2^{10} - C(10,0) - C(10,1) - C(10,2) = 968$ There are 968 polygons.

13.   Use the methods of Pascal's Triangle.

1		3		3		2
	1		2		X	
		X		2		1
			1		1	
				☻		

$1 + 3 + 3 + 2 = 9$
There are 9 paths to the top.

14. (a) $(3x + 4)^6 = C(6,0)(3x)^6 + C(6,1)(3x)^5(4) + C(6,2)(3x)^4(4)^2 +$

$C(6,3)(3x)^3(4)^3 + C(6,4)(3x)^2(4)^4 + C(6,5)(3x)(4)^5 + C(6,6)(4)^6$

$= 729x^6 + 5832x^5 + 19440x^4 + 34560x^3 + 34560x^2 + 18432x + 4096$

(b) $\left(m^2 + \dfrac{2}{m}\right)^4 = C(4,0)(m^2)^4 + C(4,1)(m^2)^3\left(\dfrac{2}{m}\right) + C(4,2)(m^2)^2\left(\dfrac{2}{m}\right)^2$

$+ C(4,3)(m^2)\left(\dfrac{2}{m}\right)^3 + C(4,4)\left(\dfrac{2}{m}\right)^4$

$= m^8 + 8m^5 + 24m^2 + 32m^{-1} + 16m^{-4}$

15. 6 terms $\quad \therefore \; n = 5$

$x^{15} = (x^3)^5$

$-1024 = (-4)^5$

$x^{15} - 20x^{12} + 160x^9 - 640x^6 + 1280x^3 - 1024 = (x^3 - 4)^5$

16. (a) $1 \times \dfrac{1}{6} = \dfrac{1}{6}$

(b) $P(4) + P(8) + P(12) = \dfrac{3}{36} + \dfrac{5}{36} + \dfrac{1}{36}$

$= \dfrac{1}{4}$

(c) $P(3) + P(6) + P(9) + \text{part (b)} = \dfrac{2}{36} + \dfrac{5}{36} + \dfrac{4}{36} + \dfrac{1}{4}$

$= \dfrac{5}{9}$

17. (a) $P = 1 - \dfrac{P(26,4)}{26^4} = 0.2148$

(b) $\dfrac{1}{4!} = \dfrac{1}{24}$

(c) $P = 1 - P(\text{no A})$

$= 1 - \dfrac{25^4}{26^4}$

$= 0.1452$

18. $P = \dfrac{0.12 \times 0.45}{0.12 \times 0.45 + 0.15 \times 0.55} = 0.3956$

19. (a) $E = \dfrac{12 \times 20}{100} = 2.4$ There are, on average, 2.4 yellow jelly beans.

(b) $P(7 \text{ yellow}) = \dfrac{C(20,7) \times C(80,5)}{C(100,12)} = 0.00177$

20. (a) $P = C(5,2)\left(\dfrac{12}{52}\right)^2 \left(\dfrac{40}{52}\right)^3 \doteq 0.24$

Odds against $\doteq 76:24$

$= 19:6$

21. (a) $P = \left(\dfrac{2}{3}\right)^4 \left(\dfrac{1}{3}\right) = 0.0658$

(b) $E = \dfrac{\dfrac{2}{3}}{\dfrac{1}{3}} = 2$ The expected waiting time is 2 green lights.

Solutions to sample examination 2

1. (a) 3 628 800   (b) 3 603 600

     (c) 36   (d) $2^{11} = 2048$

2. $6^5 = 7776$ $\therefore$ 7776 outcomes are possible.

3. (a) $8! = 40\ 320$ The cards could be arranged in 40 320 ways.

     (b) $1 \times 7! = 5040$ The cards could be arranged in 5040 ways.

4. $\dfrac{12!}{3!5!2!2!} = 41\ 580$ It could be done in 41 580 ways.

5. Total in a circle minus opposite $= 1 \times 7! - 1 \times 6!$

                                              $= 4320$

     They could be arranged in 4320 ways.

6. (a) $1 \times \dfrac{6!}{3!} = 120$ They could be arranged in 120 ways.

     (b) $\dfrac{7!}{3!3!} = 140$ They could be arranged in 140 ways.

     (c) Treat NU as one letter. NU $\neq$ UN, so x 2

         $\dfrac{6! \times 2}{3!2!} = 120$ They could be arranged in 120 ways.

7. $C(13,7) = 1716$ The books could be chosen in 1716 ways.

8. $C(12,3) \times C(9,3) \times C(6,3) \times C(3,3) = 369\ 600$

                                      They could be distributed in 369 600 ways.

9. $4158 = 2 \times 3 \times 3 \times 3 \times 7 \times 11$

$4 \times 2^3 = 32 \qquad \therefore 4158$ has 32 factors.

10. $5 \times 6 \times 4 - 1 = 119 \qquad$ At least 1 ball can be selected in 119 ways.

11. 2 from the top and 1 from the side or 1 from the top and 2 from the side.

$C(6,2) \times C(4,1) + C(6,1) \times C(4,2) = 96$

A triangle could be formed in 96 ways.

12. $C(n,2) = 2C(n-2,2) - 24$

$$\frac{n!}{(n-2)!2!} = \frac{2(n-2)!}{(n-4)2!} - 24$$

$n(n-1) = 2(n-2)(n-3) - 48 \qquad$ Cancel and multiply through by 2.

$n^2 - n = 2n^2 - 10n + 12 - 48$

$n^2 - 9n - 36 = 0$

$(n-12)(n+3) = 0$

$n = 12$ or $n = -3$

$n \in N \; \therefore \; n \neq -3$

$\therefore \; n = 12$

13. (a) $(4a + b)^5 = C(5,0)(4a)^5 + C(5,1)(4a)^4b + C(5,2)(4a)^3b^2$
$+ C(5,3)(4a)^2b^3 + C(5,4)(4a)b^4 + C(5,5)b^5$
$= 1024a^5 + 1280a^4b + 640a^3b^2 + 160a^2b^3 + 20ab^4 + b^5$

(b) $\left(y - \dfrac{-2}{y^2}\right)^6 = C(6,0)y^6 + C(6,1)y^5\left(\dfrac{-2}{y^2}\right) + C(6,2)y^4\left(\dfrac{-2}{y^2}\right)^2 + C(6,3)y^3\left(\dfrac{-2}{y^2}\right)^3$

$+ C(6,4)y^2\left(\dfrac{-2}{y^2}\right)^4 + C(6,5)y\left(\dfrac{-2}{y^2}\right)^5 + C(6,6)\left(\dfrac{-2}{y^2}\right)^6$

$= y^6 - 12y^3 + 60 - 160y^{-3} + 240y^{-6} - 192y^{-9} + 64y^{-12}$

13. 5 terms $\therefore \; n = 4$

$x^8 = (x^2)^4$

$625y^4 = (-5y)^4$

$x^8 - 20x^6y + 150x^4y^2 - 500x^2y^3 + 625y^4 = (x^2 - 5y)^4$

15.

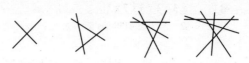

No. of stickts	2	3	4	5
No. of intersections	1	3	6	10

```
                    1
                 1     1
              1     2     1
           1     3     3     1
        1     4     6     4     1
     1     5    10    10     5     1
```

$C(2,2) = 1$
$C(3,2) = 3$
$C(4,2) = 6$
$C(5,2) = 10$
The total number of
intersections is $C(n,2)$

16. (a) $P = C(3,2)\left(\dfrac{3}{7}\right)^2 \left(\dfrac{4}{7}\right)^3 = 0.3149$

(b) $P = \dfrac{C(7,2)C(4,1)}{C(7,3)} = 0.3429$

17. (a) $P = \dfrac{1}{2001}$

(b) $P = \left(\dfrac{1}{2001}\right)^2 = \dfrac{1}{4\ 004\ 001}$

18. $p = 1 - \dfrac{P(12,5)}{12^5}$ $q = \dfrac{P(12,5)}{12^5}$

Odds against $= 95\ 040:153\ 792$
$= 55:89$

19. (a) $P = P(2) + P(3) + p(5) + p(7) + p(11)$

$= \dfrac{1}{36} + \dfrac{2}{36} + \dfrac{4}{36} + \dfrac{6}{36} + \dfrac{2}{36} = \dfrac{5}{12}$

(b) $P = \dfrac{2}{6} \times \dfrac{2}{6} = \dfrac{1}{9}$

(c) $P = \dfrac{5}{12} + \dfrac{1}{2} - \dfrac{1}{36} = \dfrac{8}{9}$

97

20. $P = \dfrac{0.75 \times 0.6}{0.75 \times 0.6 + 0.48 \times 0.4} = 0.7009$

21.   (a) $E = \dfrac{\dfrac{68}{80}}{\dfrac{12}{80}} = 5.7 = 6$

The fans should not expect a win until their seventh game.

(b) $P = \left(\dfrac{68}{80}\right)^4 \left(\dfrac{12}{80}\right) = 0.078$

22.   (a) $P = \dfrac{C(5,2)C(7,4)}{C(12,6)} = 0.3788$

(b) $E = \dfrac{6 \times 5}{12} = 2.5$ On average, there will be 2.5 boys.

23.   (a) $1000 winnings = 10 successes.
$P = C(15,10)(0.8)^{10}(0.2)^5 = 0.1032$

(b) $E = \$100 \times 15 \times 0.8 = \$1200$

On average, she would expect to win $1200.

Study tips for mathematics

- During the school year, mark off problem areas that need extra attention.

- Read over your notes. Write a summary for each topic.

- Try the review exercises in your textbook and this Coles Notes.

- Retry class exercises and tests, especially those questions where you made errors.

- Study with a friend of similar ability so that you can help each other.

- Take plenty of breaks (10-15 minutes). Get lots of sleep.

- Do not make the day of the exam a major study session. Just a quick review should be all that is necessary.

- Try a sample exam at the end of your study session, with your books closed. Give yourself a 2-hour time span with no distractions.

- You may wish to try one of the following techniques:

TECHNIQUE 1:

       Session 1   - Study Chapter 1.
       Session 2   - Quick review of Chapter 1.
                     - Study Chapter 2.
       Session 3   - Quick review of Chapters 1 and 2.
                     - Study Chapter 3

- Continue this pattern until the whole course is finished.
- Do a final review of all chapters.
- Try a cumulative review exercise.
- Try a sample final exam.

TECHNIQUE 2:

- Do a brief review of all your notes
- Go back and review each chapter. Make a 1 to 2 page summary for each chapter.
- Do selective review questions, ranging from easy to difficult questions.
- On the final study session (the day before the exam), go over each chapter summary, then try a cumulative review exercise and then a sample exam.

Important formulas

- Number of Arrangements with Repetition Permitted

  n^r

- Number of Permutations with Repetition Not Permitted

  $P(n,r) = \dfrac{n!}{(n-r)}$

- Number of Arrangements with Some Elements Alike

  $\dfrac{n!}{p!q!r!...}$

- Number of Combinations of r Elements

  $C(n,r) = \dfrac{n!}{(n-r)!r!}$

- Total Number of Subsets of a Set 2^n

- Total Number of Subsets with Some Elements Alike

  $(p+1)(q+1)(r+1)...2^{n-p-q-r}$

- Binomial Expansion

  $(x+y)^n = C(n,0)x^n +$
  $C(n,1)x^{n-1}y + C(n,2)x^{n-2}y^2$
  $+ ... + C(n,n)y^n$

- Probability of Event A

  $\dfrac{n(A)}{n(S)}$

- Expectation

  $\sum xP(x)$

- Hypergeometric Probability

  $\dfrac{C(a,x)C(b,r-x)}{C(n,r)}$

- Hypergeometric Expectation

  $\dfrac{ar}{n}$

- Binomial Probability

  $C(n,r)p^r q^{n-r}$

- Binomial Expectation

  np

- Geometric Probability

  $q^x p$

- Geometric Expectation

  $\dfrac{q}{p}$

Writing a
mathematics exam

- Look over entire exam. Budget your time based on the number of questions and the time allotted. Give yourself 10 minutes at the end for correction of errors.

- Identify the easiest questions. These should be done first.

- Look for clues in the wording of the problem that suggest model solutions.

- Frequently, more than one method is available. Use the one that is appropriate, based on the instructions.

- Often your teacher accepts more than one type of solution. If so, choose the easiest method.

- Define all variables, including appropriate quantities and units of measurement. (These are typically known as "Let" statements.)

- Write down the known values of variables before substituting them into the equations.

- Always show all steps in a solution. Teachers often give part marks if errors are made but the solution makes sense.

- Provide a concluding statement in proper sentence form. Include appropriate units of measure.

- Check: Does each answer make sense? Too often, final answers have no correct correlation with size, quantity, etc.

- Go over all your solutions, looking for errors. Many errors are careless mistakes, but some may also be misinterpretations of questions.

For fifty years, Coles Notes have been helping
students get through high school and university.
New Coles Notes will help get you through the rest of life.

NOTES & UPDATES